Springer Proceedings in Energy

More information about this series at http://www.springer.com/series/13370

Sachin Kumar · Rajesh K. Sani
Y.K. Yadav
Editors

Conference Proceedings of the Second International Conference on Recent Advances in Bioenergy Research

ICRABR 2016

Springer

Editors
Sachin Kumar
SSS National Institute of Bio-Energy
Kapurthala, Punjab
India

Y.K. Yadav
SSS National Institute of Bio-Energy
Kapurthala, Punjab
India

Rajesh K. Sani
South Dakota School of Mines
and Technology
Rapid City, South Dakota
USA

ISSN 2352-2534 ISSN 2352-2542 (electronic)
Springer Proceedings in Energy
ISBN 978-981-10-6106-6 ISBN 978-981-10-6107-3 (eBook)
https://doi.org/10.1007/978-981-10-6107-3

Library of Congress Control Number: 2017952920

Printed on acid-free paper

This Springer imprint is published by Springer Nature
The registered company is Springer Nature Singapore Pte Ltd.
The registered company address is: 152 Beach Road, #21-01/04 Gateway East, Singapore 189721, Singapore

Contents

About the Editors

Dr. Sachin Kumar has been working as a Deputy Director in the Biochemical Conversion Division at the Sardar Swaran Singh National Institute of Bio-Energy, Kapurthala, India. He completed his Ph.D. in Chemical Engineering from Indian Institute of Technology Roorkee, India. He has more than 12 years of research experience in Biochemical Conversion of Biomass to Biofuels including lignocellulosic ethanol, biogas, biohydrogen. During his career at SSS-NIBE, he has developed the state-of-the-art laboratory facilities for Biochemical Conversion of Biomass to Biofuels. He has completed three research projects and one consultancy project and actively engaged in two ongoing research projects. He is coordinating for one of the projects under Indo-Brazil bilateral collaboration from Indian side. His area of research is biofuels including bioethanol, biogas, biobutanol, biohydrogen, algal biomass, bioprocess engineering, enzyme technology, metabolic engineering. Dr. Sachin has published more than 50 papers in peer-reviewed journals, chapters and papers in conference proceedings, and 07 edited books. He has been granted one US patent and one filed patent in India. He is a recipient of 2016 ASM-IUSSTF Indo-US Research Professorship and selected as Bioenergy-Awards for Cutting Edge Research (B-ACER) Fellow 2016 by DBT and IUSSTF. He is serving as an editor, associate editor, and editorial board member of various peer-reviewed journals. He has coordinated three national and two international conferences and five national training programs.

Dr. Rajesh Sani is an Associate Professor in the Departments of Chemical and Biological Engineering and Applied Biological Sciences at South Dakota School of Mines and Technology, South Dakota. His research includes extremophilic bioprocessing of lignocellulose-based renewables for biofuels and bioproducts and bioprospecting of extremophilic microorganisms for developing more efficient and cost-effective biofuel (bioenergy) production technologies. Over the past 11 years, he has been the PI or co-PI on over $12 million in funded research. Several of his accomplishments in research and advising include the following: (i) postdocs supervised (7); (ii) graduate students supervised (MS students, 10, and Ph.D., 6), and (iii) undergraduate students and K12 teachers supervised (over 35). He has one patent and five invention disclosures, has published over 55 peer-reviewed articles in high-impact-factor journals, and has contributed several chapters. He is currently acting as editor and co-editor for three textbooks which will be published by Springer International Publishing AG. In addition, he has been a proposal reviewer and panelist for the Federal Agencies: (i) National Science Foundation, (ii) US Army Research Office, (iii) Department of Energy, (iv) US Geological Survey, and (v) User Facility—Environmental Molecular Sciences Laboratory. He also serves the Industrial Microbiology profession as "Biocatalysis Program Committee Member" of the Society for Industrial Microbiology and Biotechnology (SIMB), technical session chair at the Annual American Institute of Chemical Engineers (AIChE) and SIMB, an associate editor.

Prof. (Dr.) Y.K. Yadav received his Ph.D. in Energy Studies from the Indian Institute of Technology Delhi, and is currently Director of the Sardar Swaran Singh National Institute of Bio-Energy, Kapurthala, India. He has been actively engaged in research, teaching, consultancy, administration, and management in the field of renewable energy for more than two decades. He developed and taught renewable energy courses to undergraduate and postgraduate students at CCS Haryana Agricultural University, Hisar, India, and established a Renewable Energy Laboratory there. He has also contributed to the establishment of the Department of Energy at Tezpur University, Assam, India.

Part I
Biochemical Conversion

Chapter 1
Evaluation of Growth and Lipid Profiles in Six Different Microalgal Strains for Biofuel Production

Kashif M. Shaikh, Asha A. Nesamma, Malik Z. Abdin and Pavan P. Jutur

Abstract Microalgae have been considered as potential feedstock to produce higher biomass and lipid content that is more suitable for biofuel production than traditional oleaginous crop plants, thus seems to be on niche of accumulating energy reserves to produce next-generation renewables such as biofuels and high-value chemicals, an essential alternative for diminishing fossil fuels. Evaluation of growth and lipid profiles of few oleaginous microalgae under nutrient deprivation will be the method to identify best industrial strain for production of biofuel precursors at commercial level. In the present study, we have evaluated six microalgal (both marine and freshwater) strains to find out their metabolic responses on growth and lipid profiles under different nutrient limitation (nitrogen, phosphorous, and/or sulfur) conditions. Our results demonstrate that all these strains showed severe growth hampering by stress phenomenon under nutrient deprivation except for phosphorous limitation, wherein the growth was normal among marine strains. Algal oils are rich in the triacylglycerols (TAGs) that serve as material for conversion to biofuels. Therefore, changes triggered by nutrient deprivation in these microalgae primarily increased TAG content ($\sim$up to 20 mg L^{-1} D^{-1}) among marine strains under nitrogen and phosphorous limitation, whereas among freshwater strains, nitrogen limitation played a major role in increasing the TAG content ($\sim$up to 15 mg L^{-1} D^{-1}). In conclusion, the biomass and lipid productivity among marine strains seems to be higher when compared to freshwater strains. Among all these six potential strains, we evaluated and identified a suitable marine strain *Parachlorella kessleri* with better biomass and higher lipid productivity for further characterization, which may be a critical step toward making algae-derived biofuels economically competitive for industrial production.

K.M. Shaikh · A.A. Nesamma · P.P. Jutur (✉)
Omics of Algae Group, Integrative Biology & DBT-ICGEB Centre for Advanced Bioenergy Research, International Centre for Genetic Engineering and Biotechnology, Aruna Asaf Ali Marg, New Delhi 110067, India
e-mail: jppavan@icgeb.res.in

K.M. Shaikh · M.Z. Abdin
Department of Biotechnology, Faculty of Science, Centre for Transgenic Plant Development, Jamia Hamdard, Hamdard Nagar, New Delhi 110062, India

S. Kumar et al. (eds.), *Conference Proceedings of the Second International Conference on Recent Advances in Bioenergy Research*, Springer Proceedings in Energy, https://doi.org/10.1007/978-981-10-6107-3_1

Keywords Biomass · Lipid · Microalgae · Oleaginous · Triacylglycerols

1.1 Introduction

Microalgal biofuels are considered as one of the viable alternative sources of energy as they are renewable, sustainable, and environmental friendly [1]. The rapid increase in global energy demand is enforcing to develop sustainable renewable resources, as the world's energy consumption is estimated to grow by 53% at the end of 2035 [2]. The production of biofuels from microalgae has gained considerable attention to be a potential feedstock capable of converting atmospheric CO_2 to substantial biomass and valuable biofuels which is of greater importance for the food and energy industries [3–6]. Microalgae are prolific photosynthetic organisms that have the potential to produce biomass along with the ability to synthesize and accumulate large quantities of lipids for biodiesel production [7, 8]. Microalgae tend to accumulate valuable compounds such as triacylglycerols (TAGs) only under stress conditions that also limit growth [9]. Thus, nutrient deprivation among microalgae provides a facile experimental tool to induce and study triacylglycerol (TAG) accumulation under different stresses on the withdrawal of nitrogen (N), phosphorus (P), and sulfur (S).

The major concerns or/and need for the production of biofuels from microalgae are mainly because fossil fuel reserves are limited and non-regenerable, and their availability will soon diminish leading to hike in fuel prices. As the fuel and food prices are closely linked, the hike in fuel prices will ultimately affect the cost of food [10]. Microalgae are the preferred option for the biofuel production due to following advantages such as:

1. rapid growth, which provides higher biomass yield;
2. high amount of lipids, which can be used to produce biodiesel [3, 11, 12];
3. use of non-arable land and a wide variety of water sources like marine/brackish waters [13–15];
4. greenhouse gas sequestration capacity, which can mitigate global warming effects [16, 17];
5. the ability to utilize nutrients from wastewaters, which are economical and environmental benefits of wastewater bioremediation [18, 19];
6. the production of valuable coproducts for commercial applications [20, 21];
7. these microalgal-based biodiesel derived fuels can be integrated into the current scenario of transportation infrastructure [22].

Although microalgal biofuels hold great promise, still considerable challenges exist for their commercialization. Several research efforts are still required to make microalgal biofuels cost-effective and sustainable, which include: selection and bioengineering microalgal strains for the best biofuel producers; optimizing culturing conditions; developing suitable bioreactors for large-scale production;

improved efficiency in biomass harvesting and downstream processing, thus reducing production costs and energy consumption [1, 23–25]. Cultivation of microalgae is influenced by various environmental factors, including physical factors, such as temperature, light, and chemical factors, such as nutrients, salinity, and pH [2, 26]. These environmental factors not only affect the biomass accumulation but also have profound influence on the biofuel productivity, thus altering the biochemical composition of the cell.

The major prerequisite for the commercial production of microalgal oil-derived biodiesel is to screen for high lipid productivity, and fast-growing algae that can fulfill the criteria for sustainable biofuels. Generally, the lipid composition among microalgae varies between 10 and 60% (dw) depending on the species as well as the environmental conditions [5, 27–29]. Recently, various biochemical strategies have been employed to improve lipid accumulation and biomass production [30, 31]. Nutrient availability has a significant impact on the growth of microalgae as well as broad effects on their lipid and fatty acid composition [32]. Henceforth, nutrient deprivation is one of the most widely used and applied lipid induction techniques followed in enhancing microalgal TAG production where they tend to change their metabolic profiles and biochemical composition. In the present study, we focussed on screening and characterization of six different microalgal (both marine and freshwater) strains depending on their growth (biomass) and lipid profiles under different nutrient limitations, viz. nitrogen, phosphorous, and sulfur for biofuel production.

1.2 Materials and Methods

1.2.1 Strains and Culture Conditions

Six microalgal (both marine and freshwater) strains were used in this study. These strains were procured from different repositories, i.e., *P. kessleri* (provided by Mr. Shrikumar Suryanaran, Indian Institute of Technology, Madras), *Dunaliella salina*, and *Nannochloropsis oculata* (procured from UTEX Culture Collection of Algae, Austin, TX USA), *Ankistrodesmus* sp., *Chlorococcum* sp., and *Desmodesmus* sp. (procured from Institute of Bioresources and Sustainable Development (IBSD), Imphal, Manipur, India). Unless stated otherwise, cultures were incubated under continuous illumination (~ 120 µmol m^{-2} s^{-1} photosynthetically active radiation [PAR]) on an orbital shaker (200 rpm) at 25 °C and ambient levels of CO_2. Cells were initially grown photoautotrophically to the middle of the logarithmic phase in F/2 medium [33] for marine strains or high salt (HS) medium [34] for freshwater strains. The composition of media components for F/2 media (g L^{-1})—NaCl—24; $MgCl_2.6H_2O$—11; Na_2SO_4—4; $CaCl_2.6H_2O$—2; KBr—0.1; H_3BO_3—0.03; $Na_2SiO_3.9H_2O$—0.005; $SrCl.6H_20$—0.04; NaF—0.003; NH_4NO_3—0.002; $Fe_3PO_4.4H_2O$—0.001; $NaNO_3$—0.075; $NaH_2PO_4.2H_2O$—0.005; $Na_2SiO_3.9H_2O$

—0.03; Vitamin B_{12}*—0.135 mg L^{-1}; Biotin vitamin solution*—0.025 mg L^{-1}; Thiamine vitamin solution*—0.335 mg L^{-1} (*add these solutions after autoclave only); and trace metals solution [in g L^{-1}]—1 ml L^{-1}; [$ZnSO_4.7H_2O$—0.023; $MnSO_4.H_2O$—0.152; $Na_2MoO_4.2H_2O$—0.007; $CoSO_4.7H_2O$—0.014; $CuCl_2.2H_2O$—0.007; $Fe(NH_4)2(SO_4)_2.6H_2O$—4.6; $Na_2EDTA.2H_2O$—4.4]; for high salt (HS) media, (mg L^{-1})—NH_4Cl—100; $MgSO_4.7H_2O$—4; $CaCl_2.2H_2O$—2; K_2HPO_4—288; KH_2PO_4—144; and Hutner's trace metals solution [in g L^{-1}]—1 ml L^{-1}, [$Na_2EDTA.2H_2O$—50; $ZnSO_4.7H_2O$—22; H_3BO_3 —11.4; $MnCl_2.4H_2O$ —5.1; $CoCl_2.6H_2O$—1.6; $CuSO_4.5H_2O$—1.16; $(NH_4)_6Mo_7O_{24}.4H_2O$—1.1; $FeSO_4.7H_2O$—5.0]. These pre-cultured cells were collected by centrifugation and resuspended at the density of 2×10^6 cells mL^{-1} in regular F/2 or HS media or in the same media lacking nitrogen (–N), phosphorus (–P), and sulfur (–S). Samples for analysis were taken immediately after resuspension (Control, 0 day) and at the time intervals of 2, 4, 6, 8, and 10 days. Culture growth was monitored by counting cells with a hemocytometer [35]. Growth rates were calculated by the following equation [36].

$$K' = \frac{\ln \frac{N_2}{N_1}}{t_2 - t_1}$$

where N_1 and N_2 equal cell counts at time 1 (t_1) and time 2 (t_2), respectively. Doubling time was also calculated once the specific growth rate was known [37].

$$\text{Doubling time} = \frac{\ln 2}{K'}$$

1.2.2 Lipid Analysis

Total lipids were extracted with a modified Bligh and Dyer [38] procedure, transmethylated and visualized as TAGs by thin-layer chromatography (TLC) and gas chromatography-mass spectrometry (GC-MS). For thin-layer chromatography (TLC) experiments, briefly $\sim 1 \times 10^8$ cells were collected by centrifugation in a glass tube with Teflon-lined screw cap, and culture media was removed by aspiration. Lipids were extracted from the cell pellet using modified Bligh and Dyer [38] method. Methanol: Chloroform (2:1, v/v) containing 0.01% butylated hydroxytoluene was added to the cell pellet and incubated for 25 °C for 30 min. After addition of $CHCl_3$ (1 mL) and water (1.8 mL), the tube was shaken vigorously and the content portioned into two phases by centrifugation at $3{,}000\times g$. The upper phase was discarded and the lower organic phase, containing the extracted lipids, was transferred to a new glass tube. Extracted lipid was dried under stream of nitrogen (N_2) and resuspended in $CHCl_3$:MeOH (100 μL, 6:1 v/v). This extract was applied to a silica 60 thin-layer chromatography plate (Sigma-Aldrich), and neutral lipids were resolved using a solvent system of 70:30:1 (v/v/v) heptane:Et_2O:AcOH. The TAG band was identified by co-migration with the TAG standard, stained highly with iodine vapors [39].

For FAME analysis, lipids in the microalgal cells (approximately $\sim 1 \times 10^8$ cells) were hydrolyzed and methylesterified in 300 μL of a 2% H_2SO_4 and methanol solution for 2 h at 80 °C. Prior to the reaction, 50 μg of heptadecanoic acid (Alfa Aesar) was added as internal standard. After esterification step, 300 μL of 0.9% (w/v) NaCl solution and 300 μL of hexane were added and mixed thoroughly for 20 s. To separate the phase, samples were centrifuged at 3,000×*g* for 3 min. A total of 1 μL of hexane layer was injected into an Agilent 6890 gas chromatograph (GC) coupled to mass spectrometer (MS) [37, 40]. The running conditions for GC-MS were described by Agilent's RTL DBWax method [41].

1.3 Results and Discussion

1.3.1 Growth Curve

Six microalgal (both marine and freshwater) strains were used in this study on the basis of standard growth and lipid profiles to determine a suitable candidate for biofuel production. Table 1.1 shows the pattern of growth among both marine and freshwater microalgal strains under standard growth conditions. The growth of microalgal strains is represented in two groups namely marine and freshwater. The fastest growing strain *Ankistrodesmus* sp. is 1.6 times faster than the slowest growing strain *Chlorococcum* sp. (Table 1.1). Among the marine strains *Parachlorella kessleri* and *Nannochloropsis oculata*, biomass production was 0.54 and 0.55 g L^{-1} and exhibited similar growth rates, achieving 0.67 and 0.69 day^{-1} with doubling times of 24.7 and 24.0 h, respectively. In freshwater strains *Ankistrodesmus* sp., biomass production was 0.32 g L^{-1}, and exhibited growth rates and doubling time, achieving 0.81 day^{-1} and 19.5 h, respectively, seem to be fastest growing than other two freshwater strains. All these strains had almost similar growth rates and doubling times with few exceptions (Table 1.1). Studies have demonstrated that nutrient-dependent lipid accumulation in microalgae under optimal growth conditions (i.e., adequate supply of nutrients including C, N, P, S, and sunlight), biomass productivity can exceed 30 g dry weight per square meter per day with very low (<5% w/w) lipid content and is also species dependent [42]. In principle, biomass production and lipid biosynthesis compete for photosynthetic assimilation of inorganic carbon, and a metabolic shift is required to switch from biomass synthesis to energy storage metabolism [43, 44]. Thus, lipids are typically believed as storage reserves within the cell that enables the organism to survive adverse environmental and/or nutrient-depleted conditions. Henceforth, the trade-offs in terms of biomass versus lipid accumulation depends upon the different levels of perturbation [45].

Table 1.1 Comparison of growth parameters in six microalgal strains under standard growth conditions

Species	Biomass production (g L^{-1})	Specific growth rate μ (day^{-1})[a]	Doubling time (h)
Parachlorella kessleri	0.54	0.67 ± 0.04	24.7
Dunaliella salina	0.48	0.62 ± 0.03	26.7
Nannochloropsis oculata	0.55	0.69 ± 0.03	24.0
Ankistrodesmus sp.	0.32	0.81 ± 0.05	19.5
Chlorococcum sp.	0.53	0.54 ± 0.04	31.0
Desmodesmus sp.	0.25	0.69 ± 0.02	24.0

[a]Average of three biological replicates ± standard deviations

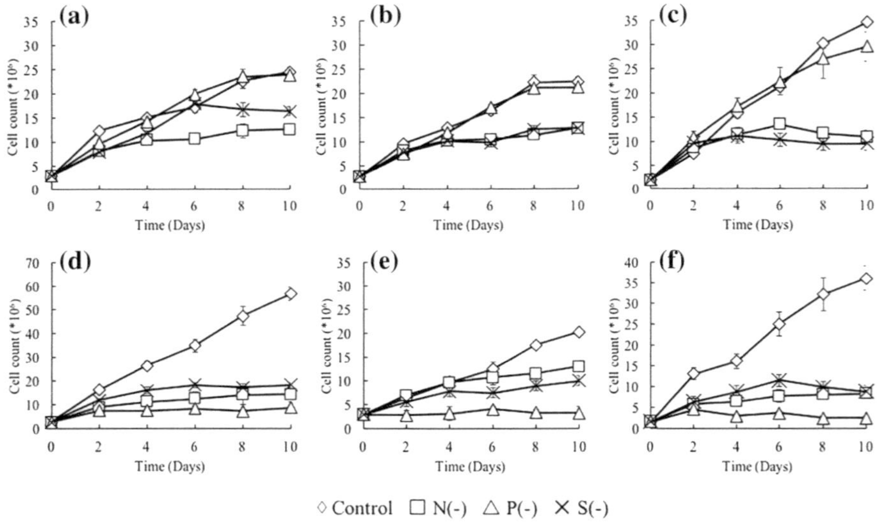

Fig. 1.1 Growth patterns of six microalgal (both marine and freshwater) strains under normal and stress condition; **a** *Parachlorella kessleri*, **b** *Dunaliella salina*, **c** *Nannochloropsis oculata*, **d** *Ankistrodesmus* sp., **e** *Chlorococcum* sp., and **f** *Desmodesmus* sp.

To analyze the growth pattern changes in different nutrient depletions such as nitrogen (–N), phosphorus (–P), and sulfur (–S), these strains were grown under photoautotrophic conditions following sampling days of 0, 2, 4, 6, 8, and 10 days. Cultures were sampled on these specific days for analyses also. Our results demonstrate that marine strains had severe effect on growth in nitrogen (–N) depletion, whereas no effect with phosphorous (–P) depletion is seen (Fig. 1.1a–c). Figure 1.1a represents that in *P. kessleri* the effect of sulfur (–S) depletion is delayed when compared to others. Among freshwater strains, all the strains showed similar effect on deletion of N, P, and S nutrients (Fig. 1.1d–f). Phosphorous (–P) depletion seems to be most severe nutrient starvation observed among the freshwater strains.

1.3.2 Lipid Analysis and Profiling

Stress-induced strategies like nutrient depletion for enhancing lipid accumulation is the most prevalent method employed in microalgae [2, 46]. This may due to two factors: (1) Lack of requisite nutrients such as N, P, and S limits the capacity and restricts cellular division; thereby to compensate, the organism will take advantage of alternative pathways for inorganic carbon fixation, and thus store these de novo fatty acids as TAGs [39]; (2) Photosynthesis and the electron transport chain produce ATP and NADPH as energy storage and electron carrier metabolites, respectively, in eukaryotic microalgae, consumed during biomass production [47]. Under normal growth conditions, the balanced ratio of reduced and oxidized metabolites is maintained; however, during nutrient deprivation due to lack of requisite nutrients, the pool of $NADP^+$ and ADP can become depleted and as the photosynthesis is mainly controlled by availability of light and cannot be completely shutoff [45]. Fatty acid synthesis consumes NADPH and ATP; therefore, increased fatty acid synthesis replenishes the pool of required electron acceptors in the form of $NADP^+$, and de novo fatty acids are stored as lipid [48].

Total lipids were extracted from all six microalgal strains and analyzed by thin-layer chromatography (TLC) as described in materials and methods, to visualize the formation of TAGs under different nutrient starvation conditions (Fig. 1.2a–f). Among the marine strains, nitrogen (–N) and phosphorous (–P) depletion played a significant role in increasing the TAG content, while sulfur (–S) depletion increased the TAG content to some extent only in *D. salina*. Phosphorous depletion among marine strains did not show any decline in the growth patterns but still lead to an increase in the TAG content. Thus, this phenomenon can be an alternative approach to enhance lipid content among specific marine strains without compromising growth. Nitrogen starvation had a severe effect on growth but also lead to increase in TAG production by onset of nutrient depletion (2nd day) (Fig. 1.2). This demonstrates that most of the metabolic responses and biochemical changes initiate immediately upon nitrogen deprivation. In marine strains, sulfur depletion has no relevance in increasing TAG content (Fig. 1.2) except in *D. salina*. In freshwater strains, *Ankistrodesmus* sp. and *Chlorococcum* sp. showed increase in TAG accumulation under nitrogen (–N) and sulfur (–S) deprivation, while *Desmodesmus* sp. accumulated TAG under sulfur deprivation only, with a very little increase under nitrogen deprivation (Fig. 1.2d–f). Under phosphorous (–P) depletion, all the fresh water strains showed a rapid arrest in cell growth with no change in the TAG content (Fig. 1.2d–f), thus predicting the relevance of phosphorus as essential nutrient for freshwater strains.

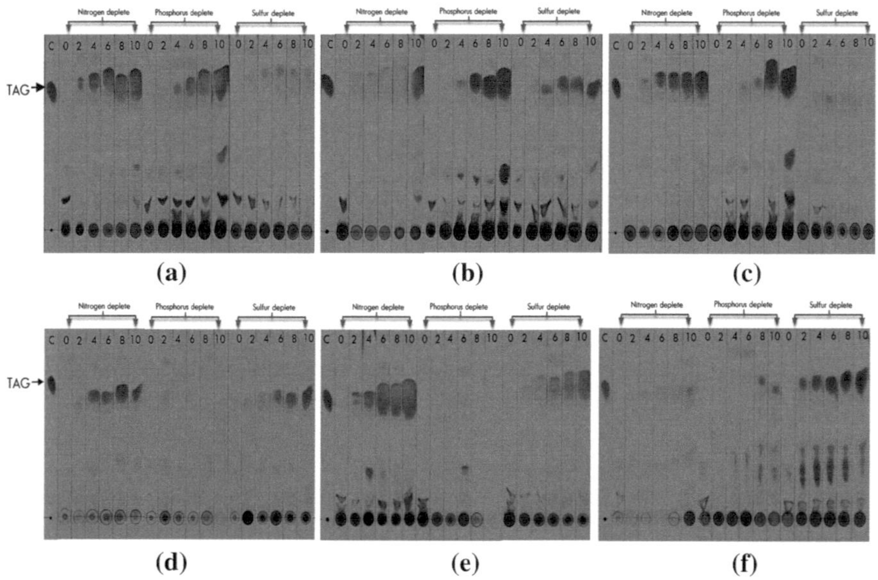

Fig. 1.2 TLC separation of neutral lipids on a silica gel G layer; **a** *Parachlorella kessleri*, **b** *Dunaliella salina*, **c** *Nannochloropsis oculata*, **d** *Ankistrodesmus* sp., **e** *Chlorococcum* sp., and **f** *Desmodesmus* sp.; C—Soy TAG (as reference for TAG mobility on TLC plate); Days of treatment—0, 2, 4, 6, 8, and 10 days; Day '0' represents TAG content on initial day of treatment (Control)

For FAME analysis, the transesterified lipids were analyzed through GC-MS as described in methodology, and the results of FAME productivity (mg L^{-1} D^{-1}) under different nutrients (N, P, and S) deprivation are illustrated in Fig. 1.3. In marine strains, FAME productivity was highest in *P. kessleri* ($\sim$20 mg L^{-1} D^{-1}) under phosphorous (–P) limitation, followed by *D. salina* ($\sim$17 mg L^{-1} D^{-1}) and *N. oculata* ($\sim$13 mg L^{-1} D^{-1}) (Fig. 1.3). Phosphorous is involved in many cellular metabolic processes and results in accumulation of lipids. In *Scenedesmus* sp. LX1, lipid accumulates up to 53% under phosphorus-limited conditions [49, 50]. The FAME productivity under nitrogen (–N) stress was highest in *P. kessleri* ($\sim$12 mg L^{-1} D^{-1}), followed by *D. salina* ($\sim$11 mg L^{-1} D^{-1}) and *N. oculata* ($\sim$7.1 mg L^{-1} D^{-1}), respectively. Nitrogen is critical for protein synthesis, but under limiting conditions, most of the carbon fixed in photosynthesis is used for producing lipid or carbohydrates, instead of proteins. Thus, nitrogen is considered to be most important nutrient affecting lipid metabolism in microalgae. For example, *Neochloris oleoabundans* and *Nannochloropsis* sp. increased about twofold and onefold, respectively, in lipid content after nitrogen deprivation [11, 51]. Interestingly, the FAME productivity among marine strains under sulfur (–S) limitation did not increase substantially as compared to other two nutritional stresses (Fig. 1.3). In freshwater strains, *Chlorococcum* sp. showed the highest FAME productivity ($\sim$12 mg L^{-1} D^{-1}) under nitrogen (–N) stress, followed by

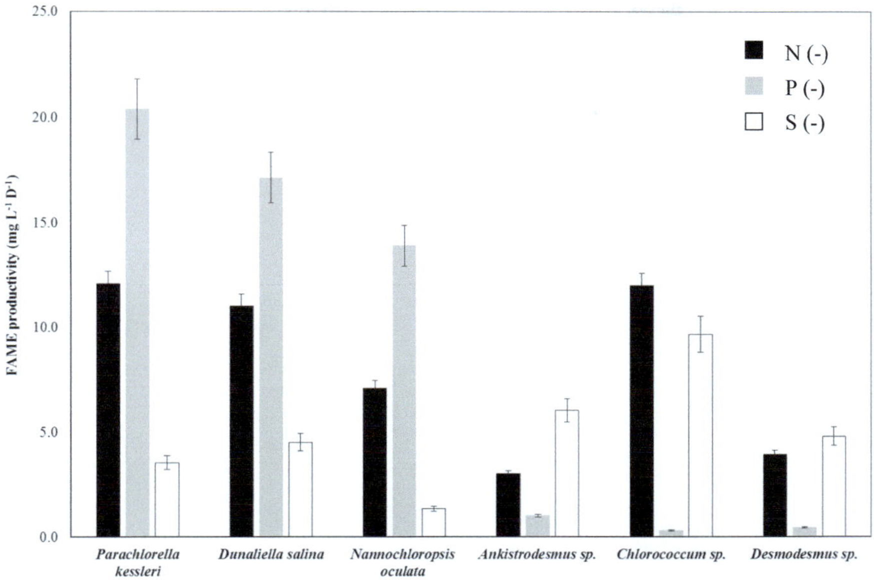

Fig. 1.3 Total FAME productivity among the six microalgal strains under nutrient-limiting conditions

Desmodesmus sp. (~7 mg L^{-1} D^{-1}), and the lowest productivity was seen in *Ankistrodesmus* sp. (~3 mg L^{-1} D^{-1}) (Fig. 1.3). Unlike marine algae, sulfur (–S) limitation in freshwater strains induced lipid production, with FAME productivity in *Chlorococcum* sp. (~10 mg L^{-1} D^{-1}), *Desmodesmus* sp. (~6 mg L^{-1} D^{-1}), and *Ankistrodesmus* sp. (~5 mg L^{-1} D^{-1}), respectively. Sulfur is one of the significant nutrients that affect the biohydrogen production in microalgae [1]. Studies have also shown that sulfur deprivation led to increased total lipid content in the green algae *Chlorella* sp. and *C. reinhardtii* [52]. Under phosphorous (–P) limitation, the cells were not able to grow and observed decreased FAME content compared to control in freshwater strains (Fig. 1.3).

Under normal conditions, FAME profile patterns among all 3 marine microalgae showed high amounts of saturated fatty acids (SFAs) and polyunsaturated fatty acids (PUFAs) with a lower proportion of monounsaturated fatty acids (MUFAs). Under –N conditions, all the marine strains exhibit an increase in MUFAs with considerably decrease in SFAs and PUFAs, probably because of the recycling of membrane lipids toward TAGs, except for *N. oculata* where PUFAs also increased with MUFAs. Though –P limitation induced highest lipid productivity, the FAME pattern suggests decrease in SFAs and considerable increase in MUFAs, but no significant increase or decrease in PUFAs. Under –S repletion, SFAs and MUFAs increased significantly with decrease in PUFAs in *P. kessleri* and *D. salina*, while in *N. oculata* SFAs increased drastically, accompanied by decrease in PUFAs with no change in MUFAs (Table 1.2). In freshwater strains, FAME profile analysis

Table 1.2 Comparison of total FAME profiles under normal versus nutrient-limiting (–N, –P, and –S) conditions after 10 days of treatment

Marine

	Parachlorella kessleri				*Dunaliella salina*				*Nannochloropsis oculata*			
	Day 10				Day 10				Day 10			
	Control	(–)N	(–)P	(–)S	Control	(–)N	(–)P	(–)S	Control	(–)N	(–)P	(–)S
Saturated fatty acid (%)	48.4	28.1	25.5	56.3	51.5	34.1	30.1	57.1	55.6	34.1	25.9	72.1
Monounsaturated fatty acid (%)	9.9	38.0	34.8	14.1	9.0	39.0	31.7	13.9	14.6	28.2	30.7	13.4
Polyunsaturated fatty acid (%)	41.6	33.9	39.7	29.6	39.6	26.8	38.2	28.9	29.8	37.7	43.4	14.6
FAME content (mg L^{-1})	13.5	120.8	203.9	35.8	16.3	110.2	171.4	45.3	8.3	70.9	139.1	13.7
FAME content (mg mg^{-1} dw)	0.19	0.42	0.37	0.08	0.23	0.38	0.34	0.15	0.17	0.27	0.20	0.06

Fresh water

	Ankistrodesmus sp.				*Chlorococcum* sp.				*Desmodesmus* sp.			
	Day 10				Day 10				Day 10			
	Control	(–)N	(–)P	(–)S	Control	(–)N	(–)P	(–)S	Control	(–)N	(–)P	(–)S
Saturated fatty acid (%)	76.5	40.7	73.6	31.3	45.8	23.5	61.0	22.3	36.4	29.9	66.7	27.6
Monounsaturated fatty acid (%)	15.1	19.7	7.6	12.0	15.2	37.6	23.1	40.3	9.0	11.9	22.0	9.8
Polyunsaturated fatty acid (%)	8.4	39.5	18.8	56.7	38.9	39.0	15.9	37.4	54.5	58.1	11.3	62.6
FAME content (mg L^{-1})	5.6	30.5	10.3	60.4	19.6	120.0	3.1	96.7	7.2	39.6	4.5	48.2
FAME content (mg mg^{-1} dw)	0.08	0.19	0.05	0.14	0.27	0.38	0.04	0.40	0.10	0.18	0.04	0.20

showed varying trend among these strains. Under normal conditions, *Ankistrodesmus* sp. showed higher amounts of SFAs (~75% w/w), whereas *Chlorococcum* sp. had SFAs and PUFAs in higher amounts (~40% w/w). Meanwhile, *Desmodesmus* sp. showed the highest content of PUFAs (~55% w/w) and SFAs (~36% w/w) (Table 1.2). Under –N limitation, *Ankistrodesmus* sp. showed a huge increase in PUFAs (from 8 to 39% w/w) accompanied by a decrease in SFAs (from 75 to 40% w/w). In *Chlorococcum* sp., SFAs decreased (from 46 to 23% w/w) accompanied by an increase in MUFAs (up to 37% w/w), while no change was observed with PUFAs. In *Desmodesmus* sp., not much change was observed under –N limitation except for a little increase in MUFAs (Table 1.2). In case of –P limitation, mostly, SFAs increased in all the freshwater strains. In sulfur (–S) limited strains, *Ankistrodesmus* sp. showed decrease in SFAs and an increase in PUFAs, while *Chlorococcum* sp. increased in MUFAs and PUFAs with a little decrease in SFAs, whereas *Desmodesmus* sp. showed a slight decrease in SFAs and MUFAs, but the PUFAs increased substantially (Table 1.2). Consequently, thus, desired biofuels precursors can be achieved by manipulating nutrient conditions for enhanced production of sustainable renewables.

1.4 Conclusions

In the present study, marine strains performed better in production of TAG and biomass compared to freshwater ones. Nutrient limitation in all the strains led to a severe growth arrest, except in marine strains under phosphorous limiting. Nitrogen (–N) starvation increased the lipid content in both freshwater and marine strains, while –P limitation played a major role in lipid accumulation in marine strains, whereas –S limitation increased lipid production in freshwater strains. Lipid profiling shows a recycling of lipids and change in the saturation and unsaturation level of fatty acids under different nutrient-limiting conditions. Although lipid increased in case of –P limitation in marine microalgal strains, the lipid profile showed a higher PUFA content, which is not desirable for biodiesel production. Nitrogen limitation increased the lipid content to higher levels, and the saturation and unsaturation levels were found to be favorable for biofuel precursors. In conclusion, the biomass and lipid productivity among marine strains seems to be higher when compared to freshwater strains. Among all these six potential strains, we evaluated and identified a suitable marine strain *P. kessleri* with better biomass and higher lipid productivity for further characterization, which may be a critical step toward making algae-derived biofuels economically competitive for industrial production.

Acknowledgements Authors acknowledge financial support from Department of Biotechnology (DBT), Ministry of Science and Technology, Govt of India. KMS was supported by a fellowship from University Grants Commission (UCG), New Delhi, India.

References

1. Cheng D, He Q (2014) Assessment of environmental stresses for enhanced microalgae biofuel production–an overview. Front Energy Res 2:1–8
2. Hu Q, Sommerfeld M, Jarvis E, Ghirardi M, Posewitz M, Seibert M, Darzins A (2008) Microalgal triacylglycerols as feedstocks for biofuel production: perspectives and advances. Plant J. 54:621–639
3. Chisti Y (2007) Biodiesel from microalgae. Biotechnol Adv 25:294–306
4. Jutur PP, Asha AN (2015) Marine microalgae: exploring the systems through an omics approach for biofuel production. In: Kim S-K, Lee CG (eds) Marine bioenergy-trends and developments. Taylor and Francis Group, pp 149–162
5. Mata TM, Martins AA, Caetano NS (2010) Microalgae for biodiesel production and other applications: a review. Renew Sust Energy Rev 14:217–232
6. Wijffels RH, Barbosa MJ (2010) An outlook on microalgal biofuels. Science 329:796–799
7. Doan TTY, Sivaloganathan B, Obbard JP (2011) Screening of marine microalgae for biodiesel feedstock. Biomass Bioenergy 35:2534–2544
8. Griffiths MJ, Harrison STL (2009) Lipid productivity as a key characteristic for choosing algal species for biodiesel production. J Appl Phycol 21:493–507
9. Tsai C-H, Warakanont J, Takeuchi T, Sears BB, Moellering ER, Benning C (2014) The protein compromised hydrolysis of triacylglycerols 7 (CHT7) acts as a repressor of cellular quiescence in *Chlamydomonas*. Proc Natl Acad Sci U S A 111:15833–15838
10. Scranton MA, Ostrand JT, Fields FJ, Mayfield SP (2015) *Chlamydomonas* as a model for biofuels and bio-products production. Plant J. 82:523–531
11. Rodolfi L, Chini Zittelli G, Bassi N, Padovani G, Biondi N, Bonini G, Tredici MR (2009) Microalgae for oil: strain selection, induction of lipid synthesis and outdoor mass cultivation in a low-cost photobioreactor. Biotechnol Bioeng 102:100–112
12. Scott SA, Davey MP, Dennis JS, Horst I, Howe CJ, Lea-Smith DJ, Smith AG (2010) Biodiesel from algae: challenges and prospects. Curr Opin Biotechnol 21:277–286
13. Costa JAV, de Morais MG (2011) The role of biochemical engineering in the production of biofuels from microalgae. Bioresour Technol 102:2–9
14. Day JG, Slocombe SP, Stanley MS (2011) Overcoming biological constraints to enable the exploitation of microalgae for biofuels. Bioresour Technol 109:245–251
15. Kliphuis AM, de Winter L, Vejrazka C, Martens DE, Janssen M, Wijffels RH (2010) Photosynthetic efficiency of *Chlorella sorokiniana* in a turbulently mixed short light-path photobioreactor. Biotechnol Prog 26:687–696
16. Ono E, Cuello JL (2007) Carbon dioxide mitigation using thermophilic cyanobacteria. Biosyst Eng 96:129–134
17. Packer M (2009) Algal capture of carbon dioxide; biomass generation as a tool for greenhouse gas mitigation with reference to New Zealand energy strategy and policy. Energy Policy 37:3428–3437
18. Markou G, Georgakakis D (2011) Cultivation of filamentous cyanobacteria (blue-green algae) in agro-industrial wastes and waste-waters: a review. Appl Energy 88:3389–3401
19. Rawat I, Ranjith Kumar R, Mutanda T, Bux F (2011) Dual role of microalgae: phycoremediation of domestic wastewater and biomass production for sustainable biofuels production. Appl Energy 88:3411–3424
20. Raja R, Hemaiswarya S, Ashok Kumar N, Sridhar S, Rengasamy R (2008) A perspective on biotechnological potential of microalgae. Crit Rev Microbiol 34:34–77
21. Spolaore P, Joannis-Cassan C, Duran E, Isambet A (2006) Commercial applications of microalgae. J Bio sci Bioeng 101:87–96
22. Fang S-C (2014) Metabolic engineering and molecular biotechnology of microalgae for fuel production. In: Pandey A, Lee DJ, Chisti Y, Soccol CR (eds) Biofuels from algae. Elsevier, Amsterdam, pp 47–65

23. Gong Y, Jiang M (2011) Biodiesel production with microalgae as feedstock: from strains to biodiesel. Biotechnol Lett 33:1269–1284
24. Greenwell HC, Laurens LML, Shields RJ, Lovitt RW, Flynn KJ (2010) Placing microalgae on the biofuels priority list: a review of the technological challenges. J R Soc Interface 7:703–726
25. Singh A, Pant D, Olsen SI, Nigam PS (2012) Key issues to consider in microalgae based biodiesel production. Energy Edu Sci Technol Part A Energy Sci Res 29:687–700
26. Singh NK, Dhar DW (2011) Microalgae as second generation biofuel. A review. Agro Sustain Dev 31:605–629
27. Amaro HM, Guedes AC, Malcata FX (2011) Advances and perspectives in using microalgae to produce biodiesel. Appl Energy 88:3402–3410
28. Cao J, Yuan H, Li B, Yang J (2014) Significance evaluation of the effects of environmental factors on the lipid accumulation of *Chlorella minutissima* UTEX 2341 under low-nutrition heterotrophic condition. Bioresour Technol 152:177–184
29. Demirbas A, Fatih DM (2011) Importance of algae oil as a source of biodiesel. Energy Convers Manage 52:163–170
30. Singh B, Guldhe A, Rawat I, Bux F (2014) Toward a sustainable approach for development of biodiesel from plant and microalgae. Renew Sustain Energy Rev 29:216–245
31. Talebi AF, Tohidfar M, Tabatabaei M, Bagheri A, Mohsenpor M, Mohtashami SK (2013) Genetic manipulation, a feasible tool to enhance unique characteristic of *Chlorella vulgaris* as a feedstock for biodiesel production. Mol Biol Rep 40:4421–4428
32. Sharma KK, Schuhmann H, Schenk PM (2012) High lipid induction in microalgae for biodiesel production. Energies 5:1532
33. Guillard RR, Ryther JH (1962) Studies of marine planktonic diatoms. I. *Cyclotella nana* Hustedt, and *Detonula confervacea* (cleve) Gran. Can J Microbiol 8:229–239
34. Sueoka N (1960) Mitotic replication of deoxyribonucleic acid in *Chlamydomonas reinhardtii.* Proc Natl Acad Sci U S A 46:83–91
35. Harris EH (1989) The *Chlamydomonas* source book: a comprehensive guide to biology and laboratory use, 1st edn. Academic Press, San Diego
36. Levasseur M, Thompson PA, Harrison PJ (1993) Physiological acclimation of marine phytoplankton to different nitrogen sources. J Phycol 29:587–595
37. Duong VT, Thomas-Hall SR, Schenk PM (2015) Growth and lipid accumulation of microalgae from fluctuating brackish and sea water locations in South East Queensland—Australia. Front Plant Sci 6:359
38. Bligh EG, Dyer WJ (1959) A rapid method of total lipid extraction and purification. Can J Biochem Phys 37:911–917
39. Msanne J, Xu D, Konda AR, Casas-Mollano JA, Awada T, Cahoon EB, Cerutti H (2012) Metabolic and gene expression changes triggered by nitrogen deprivation in the photoautotrophically grown microalgae *Chlamydomonas reinhardtii* and *Coccomyxa* sp. C-169. Phytochemistry 75:50–59
40. Lim DK, Garg S, Timmins M, Zhang ES, Thomas-Hall SR, Schuhmann H, Li Y, Schenk PM (2012) Isolation and evaluation of oil-producing microalgae from subtropical coastal and brackish waters. PLoS ONE 7:40751
41. Brown MR (1991) The amino acid and sugar composition of sixteen species of microalgae used in mariculture. J Exp Mar Biol Ecol 145:79–99
42. Gordon JM, Polle JE (2007) Ultrahigh bioproductivity from algae. Appl Microbiol Biotechnol 76:969–975
43. Schuhmann H, Lim DKY, Schenk PM (2012) Perspectives on metabolic engineering for increased lipid contents in microalgae. Biofuels 3:71–86
44. Valenzuela J, Mazurie A, Carlson RP, Gerlach R, Cooksey KE, Peyton BM, Fields MW (2012) Potential role of multiple carbon fixation pathways during lipid accumulation in *Phaeodactylum tricornutum*. Biotechnol Biofuels 5:1–17

45. Fields MW, Hise A, Lohman EJ, Bell T, Gardner RD, Corredor L, Gerlach R (2014) Sources and resources: importance of nutrients, resource allocation, and ecology in microalgal cultivation for lipid accumulation. Appl Microbiol Biot 98:4805–4816
46. Gao Y, Yang M, Wang C (2013) Nutrient deprivation enhances lipid content in marine microalgae. Bioresour Technol 147:484–491
47. Halsey KH, O'Malley RT, Graff JR, Milligan AJ, Behrenfeld MJ (2013) A common partitioning strategy fro photosynthetic products in evolutionary distinct phytoplankton species. New Phytol 198:1030–1038
48. Brown AP, Slabas AR, Rafferty JB (2009) Fatty acid biosynthesis in plants-metabolic pathways, structure and organization. In: Govindjee, Wada H, Murata N (eds) Lipids in photosynthesis: essential and regulatory functions, 30th edn., pp 11–34
49. Chu FF, Chu PN, Shen XF, Lam PK, Zeng RJ (2014) Effect of phosphorus on biodiesel production from *Scenedesmus obliquus* under nitrogen-deficiency stress. Bioresour Technol 152:241–246
50. Xin L, Hong-ying H, Ke G, Ying-xue S (2010) Effects of different nitrogen and phosphorus concentrations on the growth, nutrient uptake, and lipid accumulation of a freshwater microalgae *Scenedesmus* sp. Bioresour Technol 101:5494–5500
51. Li Y, Horsman M, Wang B, Wu N, Lan CQ (2008) Effects of nitrogen sources on cell growth and lipid accumulation of green alga *Neochloris oleoabundans*. Appl Microbiol Biotechnol 81:629–636
52. Matthew T, Zhou W, Rupprecht J, Lim L, Thomas-Hall SR, Doebbe A et al (2009) The metabolome of *Chlamydomonas reinhardtii* following induction of anaerobic H_2 production by sulfur depletion. J Biol Chem 284:13415–13425

Part II
Chemical Conversion

Chapter 2
Physicochemical Assessment of Pumpkin (*Cucurbita pepo* L.) Seed Oil as a Viable Feedstock for Biodiesel Production

Bichitra Bikash, Nabajit Dev Choudhury, Dilip Kumar Bora and Kalyan Kalita

Abstract Biodiesel is a renewable and sustainable alternative to fossil fuels which is derived from vegetable oils and animal fats. This paper presents the prospect of Pumpkin (*Cucurbita pepo* L.) seed oil, as a feedstock for biodiesel production because of its suitability for production in a variety of atmospheric condition, easy cultivation, high-fruit production rate, and low cost of cultivation. In this study, *C. pepo* L. seeds were collected, processed, and oil was extracted from the seeds by mechanical extraction process. The important physical and chemical properties of the extracted seed oil were experimentally analyzed. Fatty acid methyl ester (FAME) of the seed oil was produced corresponding to ASTM standards. The quality of biodiesel produced was assessed in terms of the physicochemical properties e.g., density, viscosity, heating value, flash point, acid value, and pour point, which are considered as the most important properties of a fuel for its application in C.I engines. The experimental findings of this study reveal that *C. pepo* L. seed can be a suitable feedstock for biodiesel production in commercial scale.

Keywords Biodiesel · *Cucurbita pepo* L. · Oil extraction · Physicochemical properties

2.1 Introduction

A vital segment of the growth of any civilization rests on the use of petroleum fuels in various fields, namely industry, power production, aviation, and transportation. However, the use of petroleum fuels needs to be reduced. The diesel burning that occurs in transport vehicles results in serious ecological changes, which include the

B. Bikash (✉) · N.D. Choudhury
Assam Down Town University, Guwahati 781026, Assam, India
e-mail: bichi1111@gmail.com

D.K. Bora · K. Kalita
Assam Engineering College, Guwahati 781013, Assam, India

S. Kumar et al. (eds.), *Conference Proceedings of the Second International Conference on Recent Advances in Bioenergy Research*, Springer Proceedings in Energy, https://doi.org/10.1007/978-981-10-6107-3_2

increase in global surface temperature (global warming), as well as changes in rainfall patterns and in the frequency of extreme weather events. In addition, the demand for sustainable alternatives for crude oil has increased because of limited fossil fuels reserves, increasing prices of world crude oil, and environmental concerns [1].

One of the feasible solutions that have got popularity during last few decades is the use of the fatty acid methyl esters, termed as biodiesels as alternative fuels of diesel [2]. Biodiesel is a renewable and sustainable alternative to fossil fuel that is derived from vegetable oils and animal fats [3]. When burned in diesel engines, biodiesels offer a drop of carbonated emissions and no sulfurized emissions [4]. However, being oxidized fuels, biodiesels emit nitrogen oxide (NOx) emissions when burnt on diesel engines [5]. Thus, biodiesel can be a potential solution to the energy needs and emission reductions. In this study, *Cucurbita pepo* L. seed, which is grown abundantly in India, is considered as a new feedstock for biodiesel production.

2.1.1 Objectives of This Study

This paper presents the prospect of that *C. pepo* L. seed oil as a promising feedstock for biodiesel production. As *C. pepo* L. seed oil is new in the field of biodiesel, to achieve this, collection of seed and extraction of vegetable oil, quantification of fatty acid methyl esters, and characterization of physicochemical properties e.g., density, viscosity, heating value, flash point, acid value, pour point were analyzed and compared corresponding to ASTM standards. The experimental findings of this study reveal that *C. pepo* L. seed can be a suitable feedstock for biodiesel production.

2.2 Materials and Methods

Scinas et al. [6] has reported that Pumpkin seeds are rich in oil, and the major physicochemical properties of pumpkin seed oil make it an attractive alternative application of the existing feedstock for biodiesel production in Greece. Hence, the Indian variety of Pumpkin can also be a potential feedstock for biodiesel production. The materials and methods applied are as follows:

2.2.1 Cucurbita pepo *L. Description*

The biggest International producers of *C. pepo* L. include the USA, Canada, Mexico, India, and China. It is a warm-weather crop that is usually planted in early

July. *C. pepo* L. or pumpkin is fruit of the Cucurbitaceae family which is commonly known as 'Pumpkin' all around the world. The growth rate of pumpkin from sowing of seed to harvesting is around 85–90 days, which is very high. Also by sowing 1–1.5 kg seeds a hectare, we can get 20–30 tonnes/hectare (Source: KISSAN Kerala Operations Centre, IIITM-K, NILA, Techno park Campus, Thiruvananthapuram).

2.2.2 Cucurbita pepo *L. Oil Seed Preparation*

As *C. pepo* L. is available in almost all the vegetable market of India, it was not much difficult in collecting the seeds. The collection of seed process was started from September 2014. The seeds were collected from the local vegetable markets of Chandrapur, Kolongpar, Mackhowa, Garchuk, etc. in Guwahati, Assam. Table 2.1 shows the amount of seed collected from different location of Assam.

Also, the seeds were collected from the hostel canteens of ADTU campus. About 1 kg of oil seed was obtained from 15 kg of *C. pepo* L. A total of 13 kg of pumpkin seeds were collected.

The seeds of *C. pepo* L. were separated from the fruit by cutting the fruit and then separating the seeds from the pulp and fibrous strands of the fruit. The seeds were then cleaned with cloth and dried evenly at room temperature of about 28–32 °C for five days. Figure 2.1 shows seed preparation steps. After drying the seeds were stored in airtight containers at room temperature.

2.2.3 *Oil Extraction*

The extraction process carried out was the mechanical extraction. It was carried out in an oil expeller machine. The pumpkin seeds are crushed inside the machine, and the vegetable oil is flowing out through the outlet tray of the machine (Fig. 2.1). The oil is filtered and collected in a beaker. From 1 kg of pumpkin seed about

Table 2.1 Amount of seed collection and the location

S. No.	Location	Geographical coordinate	Amount in kg
1	Chandrapur	26° 13′ 45″N 91° 55′ 10″E	2
2	Kolongpar	26° 11′ 44″N 91° 50′ 57″E	3
3	Fancy Bazar	26° 13′ 52″N 91° 44′ 23″E	6
4	Gorchuck	26° 6′ 57″N 91° 42′ 31″E	2

Fig. 2.1 Flow chart of *Cucurbita pepo* L. seed preparation to oil extraction

480 gm of seed oil is extracted. The oil is filtered and collected. From 7 kg of pumpkin seeds about 3.2 l of pumpkin seed oil was extracted. The extraction process was carried out in an oil mill at Chandrapur (26° 13′ 45″N–91° 55′ 10″E) which is nearby the University campus. Figure 2.1 shows the seed preparation and oil extraction steps.

2.2.4 Cucurbita pepo L. Seed Oil Analysis

Table 2.2 shows the important properties of *C. pepo* L. seed oil.

The experiments were performed corresponding to ASTM standards. The oil content of *C. pepo* L. seed was found to be 48%. The method of finding oil content of the seed is given below:

Table 2.2 Properties of *Cucurbita pepo* L. seed oil

S. No.	Properties	*Cucurbita pepo* L. seed oil
1	Oil content	48%
2	Density (28 °C)	800 kg m^{-3}
3	Kinematic viscosity (40 °C)	35.6 mm^2 s^{-1}
4	FFA	0.88 mg g^{-1}
5	Pour point	−12 °C
6	Water content	598 μg g^{-1}

$$\text{Oil content } (\%) = (W_2 - W_1)/W \times 100$$

where, W = weight of sample; W_1 = weight of beaker with glass ball; W_2 = weight of beaker with glass ball and oil; and $W_2 - W_1$ = weight of the oil.

Density is the mass per unit volume of the oil. The density of *C. pepo* L. found out to be 800 kg m^{-3} at 28 °C.

Kinematic viscosity is the property of a fluid which resists the fluid to flow over a surface. Higher the viscosity of a fluid greater will be the resistance to flow. Generally, a vegetable oil is more viscous than diesel. The viscosity of *C. pepo* L. found out to be 35.6 mm^2 s^{-1}.

FFA is the estimated amount KOH required to neutralize the soap formation. Higher the FFA value greater will be the soap formation during transesterification of the vegetable oil. The FFA of the *C. pepo* L. was found out to be 0.88 mg of KOH g^{-1}.

2.2.5 *Biodiesel Production*

Transesterification of was carried out in the chemistry lab of Assam down town University. Potassium hydroxide is taken as a catalyst in the reaction. The vegetable oil was first heated to 100 °C to remove the water content in the vegetable oil. Heated oil is allowed to cooled to 65 °C and put upon a magnetic stirrer. The alcohol which is methanol along with catalyst is poured in the heated oil. The catalyst used is potassium hydroxide (KOH). The molar ratio of methanol to oil is maintained at 6:1. The amount of catalyst i.e., KOH is determined by titration process. The mixture of the bio-oil and alcohol is kept in the magnetic stirrer for one hour keeping the temperature at 65 °C. The process flow for biodiesel production is shown in Fig. 2.2.

2.2.6 *Products of Transesterification*

Transesterification process has two products. Major product is biodiesel and coproduct is glycerol. After stirring for an hour the mixture is kept in a cool place for gravity separation. Two phases of glycerol and biodiesel are appeared after 24 h in the container. Glycerol is appeared in lower phase and the biodiesel floats over glycerol due to low density of the biodiesel. The biodiesel and glycerol were separated by using a separating funnel.

After separation, the biodiesel was washed by adding 50% volume of warm distilled water and agitated gently for 5 min. The mixture was allowed to settle, and this gave a two-layer mixture from which the biodiesel was separated. The process

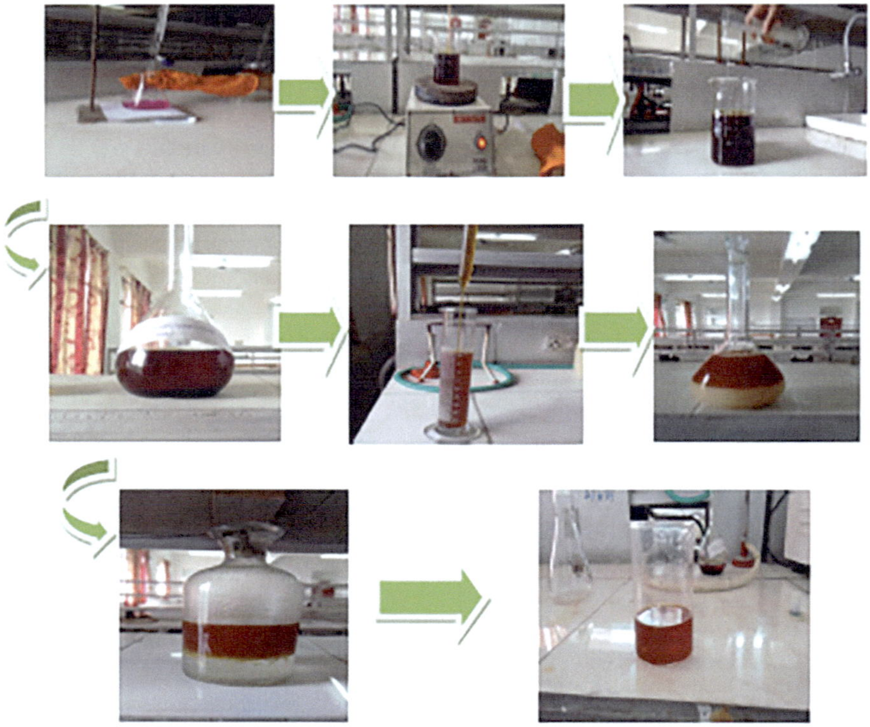

Fig. 2.2 The process flow of biodiesel production from *Cucurbita pepo* L. seed oil

was repeated two more times to give a clearer ester. The washing was done to remove impurities and any remaining residual methanol.

NMR analysis of FAME was analyzed by ^{1}H NMR and ^{13}C NMR. The ^{1}H and ^{13}C NMR spectra were recorded at 600 and 150 MHz, respectively, using Bruker Avance III 600 MHz NMR spectrometer using $CDCl_3$ as solvent. The ^{1}H and ^{13}C NMR spectra of FAME from *C. pepo* L. seed oil are depicted in Figs. 2.3 and 2.4.

The ^{1}H NMR spectrum of FAME from *C. pepo* L. seed oil is shown in Fig. 2.3. The multiplet at δ 5.240–5.398 ppm represents the olefinic protons (–CH=CH–). The triplet near ~δ 2.763 ppm indicates the bis-allylic protons (–C=C–CH_2–C=C–) of the unsaturated fatty acid (like linoleic acid) [7]. The triplet at δ 2.290 ppm represents the α-methylene protons to ester (–CH_2–CO_2Me). The α-methylene protons to double bond (–CH_2–C=C–) appear as a multiplet near ~δ 1.988–2.060 ppm. The β-methylene protons to ester (CH_2–C–CO_2Me) also appear as a multiplet at δ 1.592–1.612 ppm. The singlet signals at δ 1.249 are expected for the protons of backbone methylenes of the long fatty acid chain. The terminal methyl protons (C–CH_3) appears as a multiplet near δ 0.85–0.88 ppm. The ^{13}C NMR spectrum of biodiesel from *C. pepo* L. seed oil is shown in Fig. 2.4. The signals at δ 173.02–173.47 ppm represent the carbonyl carbon of the ester molecules of FAME, and the olefinic carbons appear at δ 128.09, 128.29, 129.58,

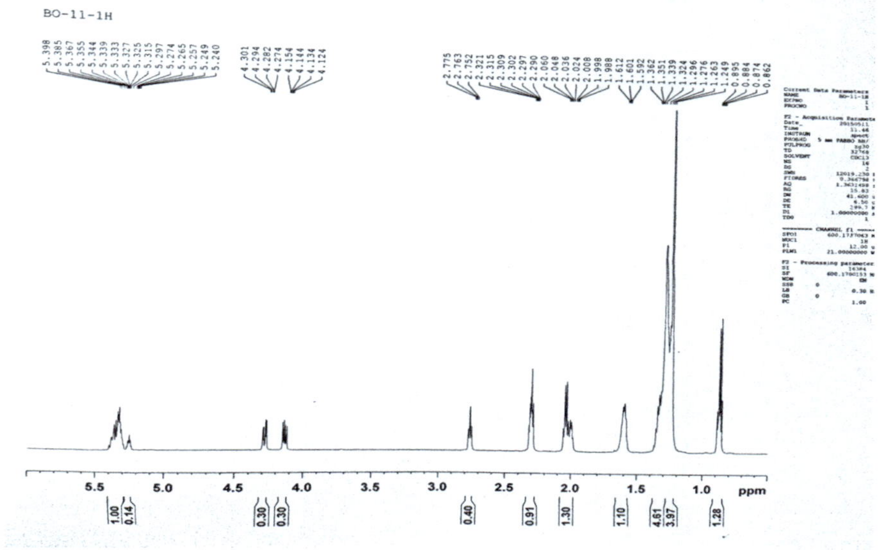

Fig. 2.3 ^{1}H NMR of FAME

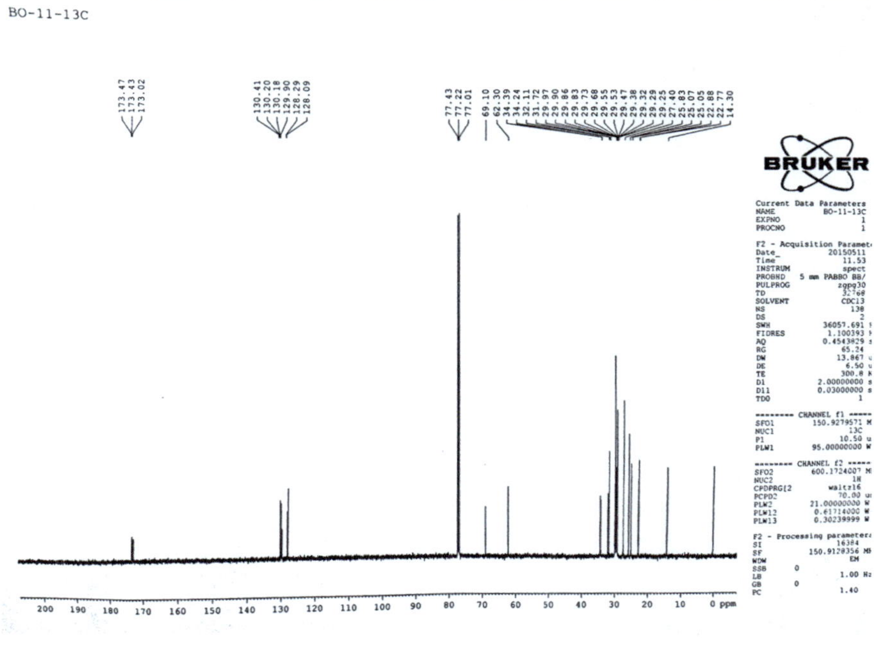

Fig. 2.4 ^{13}C NMR of FAME

129.90, 130.18, 130.20, and 130.41 ppm. The methylene and methyl carbons of fatty acid moiety appear in the range from δ 14.30 to 34.39 ppm.

2.3 Result and Discussion

2.3.1 Biodiesel Analysis

The chemical composition of biodiesels is very important for determining their suitability for diesel engine application. Chemically, all biodiesels are mono alkyl esters of fatty acids, commonly referred to as fatty acid methyl or ethyl esters [8]. The analysis of the physicochemical properties are shown in Table 2.3.

Density is the mass per unit volume of the oil. The density of *C. pepo* L. methyl ester was found out at chemistry laboratory of Assam down town University by bottle method. The density was found out to be 787 kg m^{-3} at 28 °C.

Kinematic viscosity is the property of a fluid which resists the fluid to flow over a surface. Higher the viscosity of a fluid greater will be the resistance to flow. The test for viscosity of *C. pepo* L. methyl ester was performed at Indian Institute of Technology, Guwahati (IITG). It was found out to be 4.41 mm^2 s^{-1}.

Flash point is the temperature at which the fuel particles get burned. Flash point of the *C. pepo* L. methyl ester was found to be 138 °C.

Calorific value is the amount of energy released from complete combustion of a specified amount of fuel. It is measured by MJ kg^{-1} of oil or fuel. The test for calorific value was performed at Indian Institute of Technology, Guwahati (IITG).

Moisture content test of *C. pepo* L. ester was carried out at Indian Institute of Technology, Guwahati. It is defined as the amount of water contained in the oil. The moisture content of *C. pepo* L. methyl ester was found out to be 598 μg g^{-1}.

The temperature at which a liquid becomes a semi solid is called pour point of the liquid. At this point, the liquid loses its flow ability and becomes stationary. The pour point of oil as low as possible to consider the oil as an IC engine fuel. The *C. pepo* L. methyl ester pour point was found out to be −13 °C.

Test for FFA level was performed at Chemistry laboratory of Assam down town University by the method of titration. The FFA level of *C. pepo* L. vegetable oil by

Table 2.3 Properties of *Cucurbita pepo* L. methyl ester

S. No.	Properties	Biodiesel
1	Density (28 °C)	787 kg m^{-3}
2	Kinematic viscosity (40 °C)	4.41 mm^2 s^{-1}
3	Flash point	138 °C
4	Calorific value	42.128 MJ kg^{-1}
5	Moisture content	530 μg g^{-1}
6	Pour point	−13 °C
7	Acid value, KOH	0.12 mg g^{-1}

Table 2.4 Physicochemical properties comparison

Properties	[b]Diesel	[a]*Cucurbita pepo* L. biodiesel	[b]Soybean biodiesel	[b]Jatropha biodiesel
Oil content	–	48%	15%	30–35%
Density (28 °C)	840 kg m^{-3}	787 kg m^{-3}	907.3 kg m^{-3}	620 kg m^{-3}
Kinematic viscosity (40 °C)	4.0 mm^2 s^{-1}	4.41 mm^2 s^{-1}	4.3745 mm^2 s^{-1}	4.9476 mm^2 s^{-1}
Calorific value	45 MJ kg^{-1}	42.128 MJ kg^{-1}	39.579 MJ kg^{-1}	39.5 MJ kg^{-1}
Pour point	−14 °C	−13 °C	1 °C	−10 °C
Cloud point	−13 °C	−6 °C	1 °C	−3 °C

[a]Experimental
[b]Literature: Kumar et al. [9]; Bwade et al. [10]

percentage was 0.88. After transesterification, the FFA level of the *C. pepo* L. methyl ester was found out to be 0.68%.

2.3.2 *Comparison with Diesel and Other Feedstock*

A comparison of the physicochemical properties of *C. pepo* L. seed oil methyl ester with petroleum diesel and other feedstock for biodiesel production is shown in Table 2.4.

2.4 Conclusion

In this study, Pumpkin (*C. pepo* L.) seed oil was considered as a feedstock for biodiesel production because of its suitability for production in a variety of atmospheric condition, easy cultivation, high fruit production rate, and easy availability in markets. Here, the physicochemical properties of *C. pepo* L. seed oil and *C. pepo* L. seed oil methyl ester was experimentally analyzed and compared with other established feedstock for biodiesel production along with petroleum diesel and found that its properties are comparable to petroleum diesel. It was found that the oil content of *C. pepo* L. seed around 48% which is higher than the Soybean and Jatropha seed. Here, we have produced biodiesel from *C. pepo* L. seed oil by transesterification process and determined the various properties like density, kinematic viscosity, calorific value, pour point, and cloud point corresponding to ASTM standard and found it slightly better than the biodiesel produced from Soybean and Jatropha seed oil. The calorific value of *C. pepo* L. seed oil methyl ester was 42.128 MJ kg^{-1}, which is almost equivalent to the calorific value of normal diesel. Since the oil content of *C. pepo* L. seed is higher than some established feedstock, and physicochemical properties are comparable to petroleum diesel hence, we can conclude that *C. pepo* L. seed oil can be a viable feedstock for production of biodiesel in India.

Acknowledgements The authors would like to express their thanks to Mr. Debarshi Baruah, Technical Officer, Centre for Energy, Indian Institute of Technology, Guwahati (IITG) for the technical and laboratory support and Mr. Neelkamal Basumatary of AdtU for his valuable contribution.

References

1. Masum BM, Masjuki HH, Kalam MA, Rizwanul Fattah IM, Palash SM, Abedin MJ (2013) Effect of ethanol–gasoline blend on NOx emission in SI engine. Renew Sustain Energ Rev 24:209–222
2. Van Gerpen J (2005) Biodiesel processing and production. Fuel Process Technol 86:1097–1107
3. Murugesan A, Umarani C, Subramanian R, Nedunchezhian N (2009) Bio-diesel as an alternative fuel for diesel engines—a review. Renew Sustain Energ Rev 13:653–662
4. Rizwanul Fattah IM, Masjuki HH, Liaquat AM, Ramli R, Kalam MA, Riazuddin VN (2013) Impact of various biodiesel fuels obtained from edible and non-edible oils on engine exhaust gas and noise emissions. Renew Sustain Energ Rev 18:552–567
5. Hoekman SK, Robbins C (2012) Review of the effects of biodiesel on NOx emissions. Fuel Process Technol 96:237–249
6. Schinas P, Karavalakis G, Davaris C, Anastopoulos G, Karons D, Zannikos F, Stournas S, Lois E (2009) Pumpkin (*Cucurbita pepo* L.) seed oil as an alternative feedstock for the production of biodiesel in Greece. Biomass Bioenerg 33:44–49
7. Moser BR, Vaughn SF (2010) Coriander seed oil methyl ester as biodiesel fuel: unique fatty acid composition and excellent oxidative stability. Biomass Bioenerg 34(4):550–556
8. Jahirul MI, Brow RJ, Senadeera W, Ashwath N, Rasul MG, Rahman MM, Hossain FM, Moghaddam L, Islam MA, O'Hara IM (2015) Physio-chemical assessment of beauty leaf (*Calophyllum inophyllum*) as second-generation biodiesel feedstock. Energ Rep 1:204–215
9. Kumar S, Singh J, Nanoti SM, Garg MO (2012) A comprehensive life cycle assessment of jatropha biodiesel production in India. Bioresour Technol 110:723–729
10. Bwade KE, Aliyu B, Kwaji AM (2013) Physicochemical properties of seed oil relevant to biodiesel production and other industrial applications. Int J Environ Eng Manag 13:72–78

Chapter 3
Performance and Emission Parameters of Single Cylinder VCR Engine Fuelled with Argemone Biodiesel and Blend with Diesel Fuel

Parmjit Singh, Sandeep Kumar Duran and Inderpreet Singh

Abstract In this, experimental study has been carried out to investigate the performance and emission characteristics of single cylinder, four strokes, water cooled, VCR engine fuelled with argemone biodiesel blends with diesel (AB10, AB20, AB30) at different load conditions (no load, part load and full load). The argemone methyl ester was derived from single step transesterification method due to low fatty acid value. The effect of different reaction parameters named as molar ratio, catalyst concentration, reaction temperature, reaction time, stirring speed on the properties of argemone biodiesel was also studied. The optimum reaction conditions for production of biodiesel were molar ratio 1:6 (oil to methanol), catalyst concentration 1.5% (w/w of oil) reaction temperature 75 °C, reaction time 2 h and stirring speed 550 rpm. The properties of argemone methyl ester were determined and compared with diesel standard. The result indicates that performance parameters such as brake thermal efficiency increased by 6–10%, and whereas the specific fuel consumption decreased by 4–6% for AB30 at full load conditions. On the other hand, emission parameters: carbon monoxides (70–90%), carbon dioxide (20–30%) and unburnt hydrocarbon (70–80%) were reduced for AB30 at full load conditions. The argemone methyl ester produced more nitric oxide (NO_x) emission by (12–20%) for AB30 at full load conditions than that of diesel fuel. From the experiment, investigation signifies that argemone biodiesel blends with diesel can be used as an alternative fuel for diesel engine without any engine modification.

P. Singh (✉) · S.K. Duran · I. Singh
Lovely Professional University, Jalandhar, India
e-mail: singhparmjit1991@gmail.com

S.K. Duran
e-mail: sandeep.17906@lpu.co.in

I. Singh
e-mail: inderpreets10@gmail.com

S. Kumar et al. (eds.), *Conference Proceedings of the Second International Conference on Recent Advances in Bioenergy Research*, Springer Proceedings in Energy, https://doi.org/10.1007/978-981-10-6107-3_3

Keywords Argemone mexicana oil · Single step transesterification
Biodiesel properties · Sodium metal catalyst · Methyl ester
Engine performance · Engine emission

3.1 Introduction

Day by day, increasing the demand for fossil fuel (petroleum) because of increased number of petroleum-based vehicles, automobiles, industrialization, transportation, mechanization and primary energy consumption [1]. The fossil fuels have limited availability resources, and rising crude oil prices has led to the search of substitute renewable fuels for ensuring energy security and environmental protection [2–4]. Nowadays, vegetable oil (edible and non-edible oil) is used as alternative fuel for diesel engine. It is renewable, biodegradable, non-toxic, environment friendly and oxygenated fuel. However, the main problems of vegetable oil as biodiesel used in CI engine are its higher viscosity, low volatility due to long chain fatty acid [5–8]. These fatty acids are produced common problems in CI engine like improper combustion, poor atomization of fuel spray and choking [9]. The viscosity of vegetable oil has been reduced by different techniques namely heating, dilution, pyrolysis, transesterification with alcohol, blending with alcohol or diesel and micro-emulsion [10–13]. Several researches [6, 14–19] were investigated the performance and the emission parameter of diesel engine fuelled with or without preheated vegetable oil. The brake thermal efficiency and brake-specific fuel consumption were slightly improved, when the preheated vegetable oil was used in CI engine. The main difficulty of straight vegetable oils is high viscosity, high density and low calorific value of oils [20]. The main objective of this experimental work is to find the optimum reaction conditions for production of biodiesel, and fuel properties of argemone methyl ester is compared with diesel and also examine the performance and emission parameter of single cylinder, four stroke VCR engine fuelled with diesel and argemone biodiesel blends with diesel AB10, AB20, AB30 under different load conditions compared with diesel fuel.

3.2 Material and Method

In this study, the argemone Mexicana seeds were purchased in the local market Hoshiarpur district of Punjab. The soxhlet apparatus was used to extract the oil from the seeds. Petroleum ether and hexane were used as a solvent during the mechanical extraction process, and oil was separating in the rotary vacuum evaporator. The catalyst sodium metal and methanol were provided in the laboratory. All experiment and laboratory works were performed in the R-D1 of SSS NIRE (Sardar Swarn Singh National Institute of Renewable Energy) Kapurthala, Punjab (Figs. 3.1 and 3.2).

Fig. 3.1 Argemone Mexicana seeds

Fig. 3.2 Argemone Mexicana seed oil

3.3 Biodiesel Setup

It consists of red keys reactor, water jacket, external heater or water bath, condenser or chiller, magnetic stirrer inside the reactor and rpm controller. The external heater was used to maintained temperature throughout the reaction by outside water heat jacket. The external chiller was provided to condense methanol vapour from the reaction mixture (Fig. 3.3).

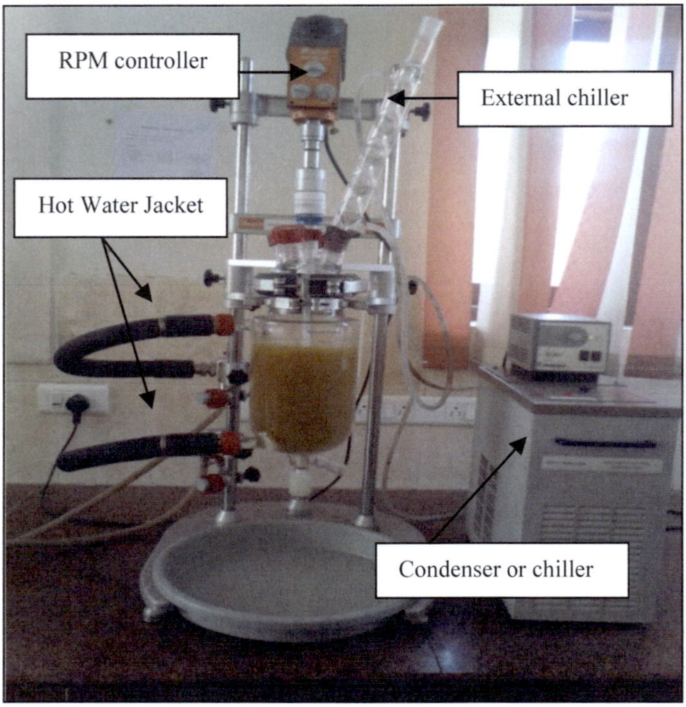

Fig. 3.3 Biodiesel setup

Table 3.1 Fatty acid composition of crude argemone oil [21]

Property	Argemone oil
Fatty acid composition (%)	
Oleic acid (C 18:1)	40.0
Linoleic acid (C 18:2)	36.6
Palmitic acid (C 16:0)	14.7
Stearic acid (C 18:0)	6.75
Palmitoleic acid (C 16:1)	1.3
Linolenic acid (C 18:3)	0.3
Arachidic acid (C 20:0)	0.3
Behenic acid (C 22:0)	0.2
Myristic acid (C 14:0)	0.1

3.3.1 Transesterification Process

The single-step transesterification was carried out in the laboratory scale setup with the molar ratio of 1:6 (oil to methanol) with base catalyst sodium metal, and catalyst concentration of 1.5% (w/w of oil) was taken. Molar ratio of argemone oil was obtained by determined the weight of fatty acid report of argemone oil. The fatty acid composition of argemone oil was shown in Table 3.1.

The quantity of methanol necessary for the transesterification process was calculated by using the molar ratio (oil to methanol) formula:

$$\frac{1}{6} = \frac{\text{Mass of oil}}{\text{Molar mass of oil}} \times \frac{\text{Molar mass of methanol}}{\text{Mass of alcohal}}$$

In this reaction process, 1000 ml oil was taken in the biodiesel preparation unit and preheat at 75 °C. The preheating of oil was used to remove unwanted moisture content. According to molar ratio 1:6 (oil to methanol), 302 ml methanol was taken in the reaction flask, and 1.5% (w/w of oil) sodium metal was used as base catalyst. About 13.65 g of sodium metal was added into the methanol reaction flask and continually stirred until the whole mixture was properly mixed. The sodium methoxide solution was poured into the reaction flask. The reaction was carried out for 2 h at constant stirring speed of 550 rpm. The transesterification reaction was over after 2 h, and whole mixture or product was put into the separating funnel. The two layers were separated for 24 h. Upper layer consists of biodiesel, and the lower layer was glycerol. The lower layer was taken out by the separating funnel so obtained argemone methyl ester. Upper layer was washed with hot water and removing some residue like excess methanol, excess catalyst and soap. During the washing process, some moisture content is remaining in the biodiesel. This moisture content was removed with the help of drying at 110 °C in oven. Thus, pure biodiesel was obtained (Figs. 3.4, 3.5 and 3.6).

Fig. 3.4 Layer separation

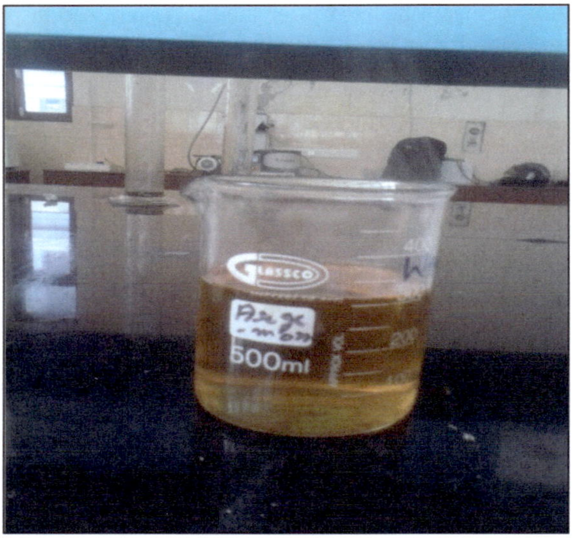

Fig. 3.5 Argemone methyl ester (biodiesel)

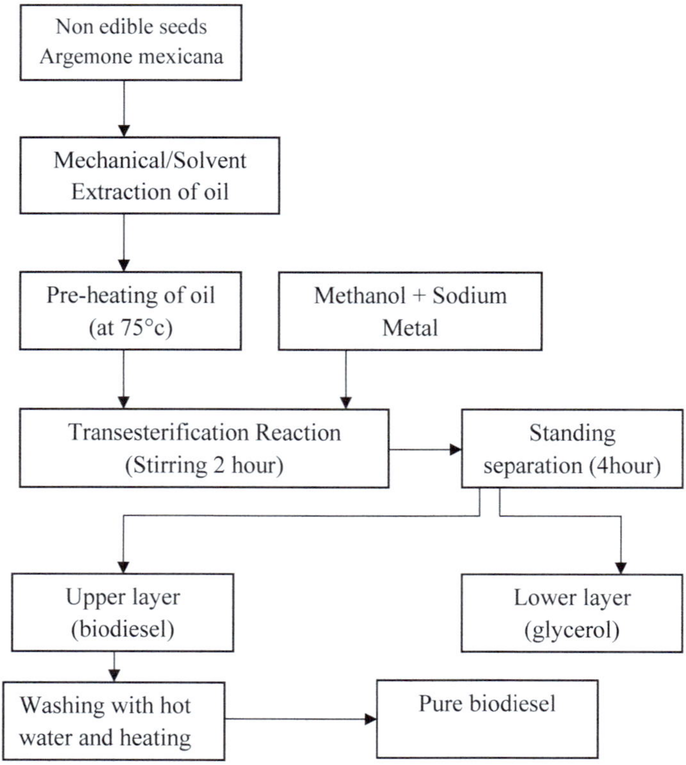

Fig. 3.6 Experiment process for production

3.4 Experimental Setup and Procedure

The experimental setup consists of single cylinder, four strokes, water cooled, variable compression ratio (VCR) diesel engine connected to eddy current type dynamometer for loading. Technical specification of the engine is shown in Table 3.2. Figure 3.7 shows the experimental engine setup. The setup enables study of VCR engine performance for brake power, indicated power, friction power, brake mean effective pressure, indicated mean effective pressure, brake thermal efficiency, indicated thermal efficiency, volumetric efficiency, mechanical efficiency, specific fuel consumption, A/F ratio and heat balance. The exhaust gas analyser model AVL DiGas 444 was used for measurement of CO, UHC, CO_2, NO_x emissions (Fig. 3.8).

3.5 Result and Discussion

The physical and chemical properties of argemone biodiesel blends with diesel (AB10, AB20 and AB30) were determined, and these properties were compared with diesel. AB10 contains 10% of biodiesel and 90% of diesel. AB20 contains 20% of biodiesel and 80% of diesel. AB30 contains 30% of biodiesel and 70% of diesel (Fig. 3.9; Table 3.3).

Table 3.2 Engine specifications

Model	TV1
Details	Single cylinder, four stroke, diesel
Bore and stroke	87.5 and 110 mm
Compression ratio	17.5:1
Cubic capacity	0.661 l
Rated power	3.5 kW @ 1500 rpm
No of cylinder	One
Cooling	Water
Inlet valve open	4.5° before TDC
Inlet valve close	35.5° after BDC
Exhaust valve open	35.5° before BDC
Exhaust valve close	4.5° after TDC
Fuel injection starts	23° before TDC
Eddy current dynamometer	Make Saj Test Plant Pvt. Ltd., Model AG10

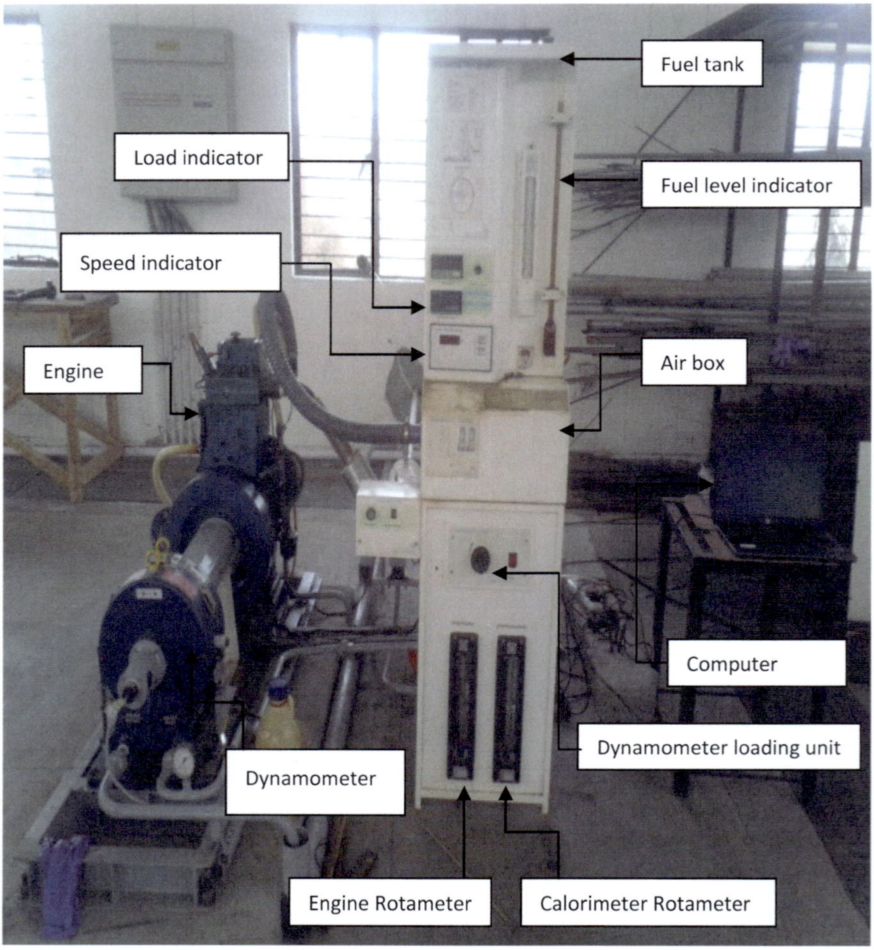

Fig. 3.7 Engine experimental setup

3.5.1 Effect of Molar Ratio on Biodiesel Yield and Kinematic Viscosity

Molar ratio is one of the main important factors effect on the biodiesel yield and kinematic viscosity of the oil. The biodiesel yield was increased with increase of oil to methanol ratio in the transesterification reaction. The molar ratio 1:6 was obtained maximum biodiesel yield of 96.67% and minimum value of kinematic viscosity 5.12 (cst). The higher value of molar ratio has a negative impact on biodiesel yield and kinematic viscosity. The biodiesel yield was reduced, and

Fig. 3.8 Exhaust gas analyzer

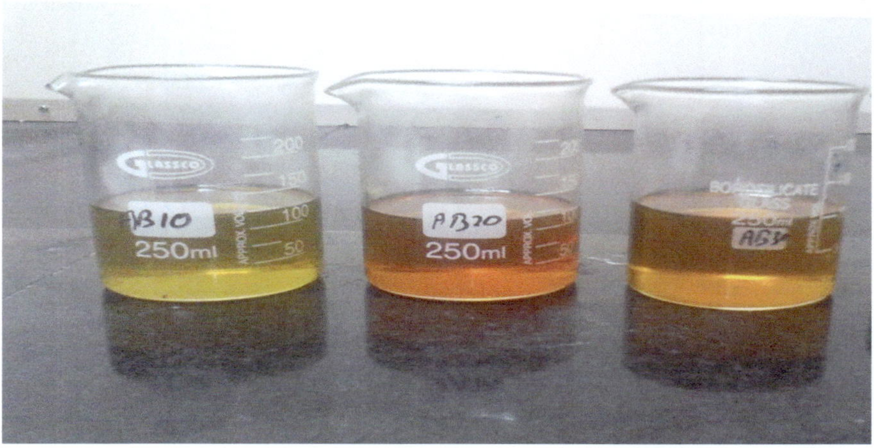

Fig. 3.9 Argemone biodiesel blends AB10, AB20, AB30

kinematic viscosity was increased for 1:8, 1:10 and 1:12 due to the excess amount of methanol in the upper layer (methyl ester) and lower layer (glycerol). The effect of molar ratio on kinematic viscosity and molar ratio was shown in Fig. 3.10.

Table 3.3 Comparison properties of diesel and argemone biodiesel blends

S. No.	Properties	Diesel	AB10	AB20	AB30
1	Density at 40 °C (g/cm^3)	0.808	0.82079	0.82678	0.83239
2	Specific gravity at 40 °C	0.814	0.83913	0.83913	0.85048
3	API density at 15 °C (g/cm^3)	0.826	0.82720	0.83824	0.83889
4	Viscosity at 40 °C (cst)	2.63	2.97	3.22	3.30
5	Flash point (°C)	65	75	83	95
6	Fire point (°C)	67	77	87	103
7	Acid value (mg KOH/g)	0.01	0.09	0.16	0.25
8	Calorific value (MJ/kg)	42	41.23	40.56	39.17

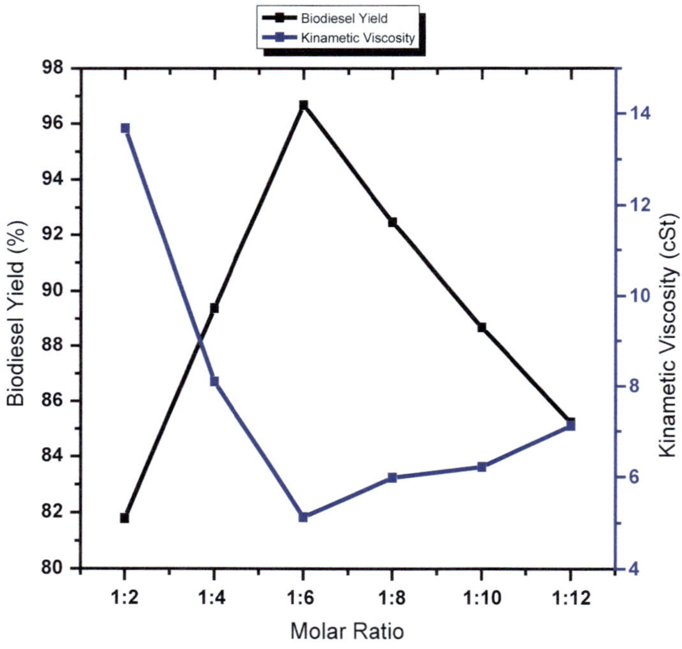

Fig. 3.10 Molar ratio versus biodiesel yield and kinematic viscosity

3.5.2 *Effect of Catalyst Concentration on Biodiesel Yield and Kinematic Viscosity*

During the reaction, different types of catalyst concentration (0.7, 0.9, 1.1, 1.3, 1.5%) were used. Higher amount of catalyst concentration was leading to more soap formation in the reaction process, which is difficult to separate methyl ester and recover glycerol. The optimum concentration of sodium methoxide was 1.5%; conversion of biodiesel was maximum of 97.12%; kinematic viscosity was

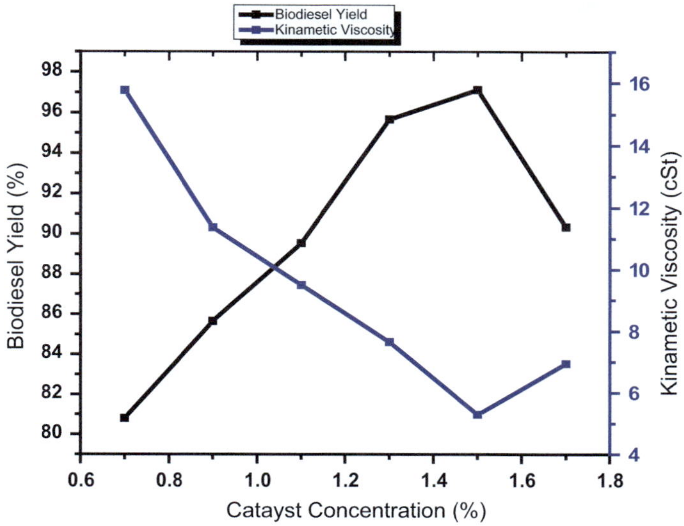

Fig. 3.11 Catalyst concentration versus biodiesel yield and kinematic viscosity

minimum of 5.29 (cst). The effect of catalyst concentration on biodiesel yield and kinematic viscosity was shown in Fig. 3.11.

3.5.3 Effect of Reaction Temperature on Biodiesel Yield and Kinematic Viscosity

The molar ratio 1:6 (oil to methanol), catalyst concentration 1.5% (w/w of oil), reaction time 2 h and stirring speed of 550 rpm were fixed. The reaction temperature was increased from 35 to 75 °C, and the production of argemone methyl ester was also increased. The reaction rate was increased more than 75 °C and resulted conversion rate was decreased, and kinematic viscosity was slightly increased due to the evaporation of methanol and saponification reaction occurs. The maximum biodiesel yield obtained at 75 °C, and kinematic viscosity was minimum at same temperature. The variation of reaction temperature versus biodiesel yield and kinematic viscosity was shown in Fig. 3.12.

3.6 Performance Parameters

The engine performance with argemone biodiesel blends AB10, AB20, AB30 was evaluated in terms of brake thermal efficiency, indicated thermal efficiency, brake mean effective pressure, indicated mean effective pressure, A/F ratio, torque, heat

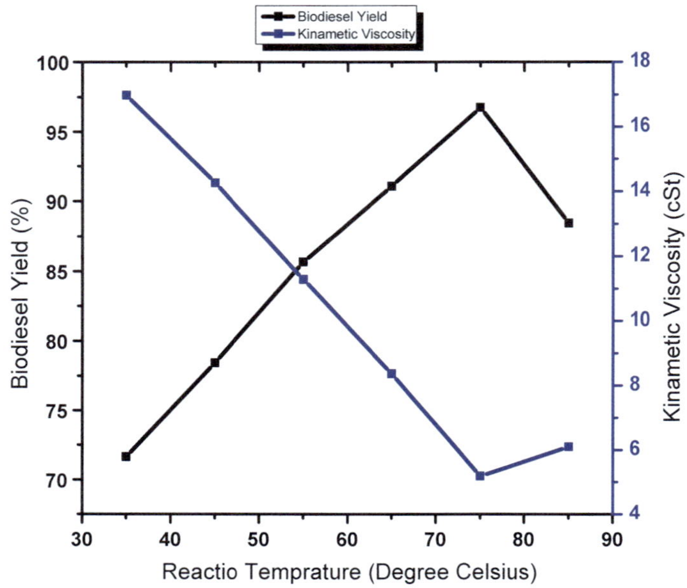

Fig. 3.12 Reaction temperature versus biodiesel yield and kinematic viscosity

loss by gas and heat loss by jacket water at different loading conditions of the engine.

3.6.1 Brake Thermal Efficiency

The variation of brake thermal efficiency versus different load conditions (no load, part load, full load) for diesel fuel and argemone biodiesel blends with diesel was shown in Fig. 3.13. The brake thermal efficiency was increased with increased in all load conditions for AB10 and AB20. Maximum brake thermal efficiency was measured at full load conditions for AB30. The brake thermal efficiency was maximum for diesel, AB10, AB20, AB30 blends of 30.58, 31.59, 31.95, 38.55% at full load conditions. The brake thermal efficiency was increased than that of diesel at full load conditions due to the less portion of friction and power loss with increasing load, and 10–12% more oxygen presence in the biodiesel fuel leads to complete combustion of biodiesel.

3.6.2 Brake Mean Effective Pressure

The variation of brake mean effective pressure versus different load conditions was shown in Fig. 3.14. The brake mean effective pressure was increased with increase in load. The brake mean effective pressure was increased due to reduction of the friction loss. The brake mean effective pressure values were 5.1 bar, 5.21 bar, 5.2 bar, 5.15 bar for diesel, AB10, AB20 and AB30 at full load conditions. The brake mean effective pressure was slightly increased than that diesel fuel due to temperature and pressure on piston increased with increase in load which in turn increasing the thermal energy releasing rate. The brake mean effective pressure is directly proportional to thermal heat release rate; therefore, BMEP also increases. The BMEP was recorded maximum for AB10 at full load conditions than that of diesel fuel.

3.6.3 Specific Fuel Consumption

The variation of specific fuel consumption with different load conditions for argemone biodiesel blends with diesel shown in Fig. 3.15. 10, 20 and 30% blend of argemone methyl ester was slightly lower than that of diesel fuel at no load

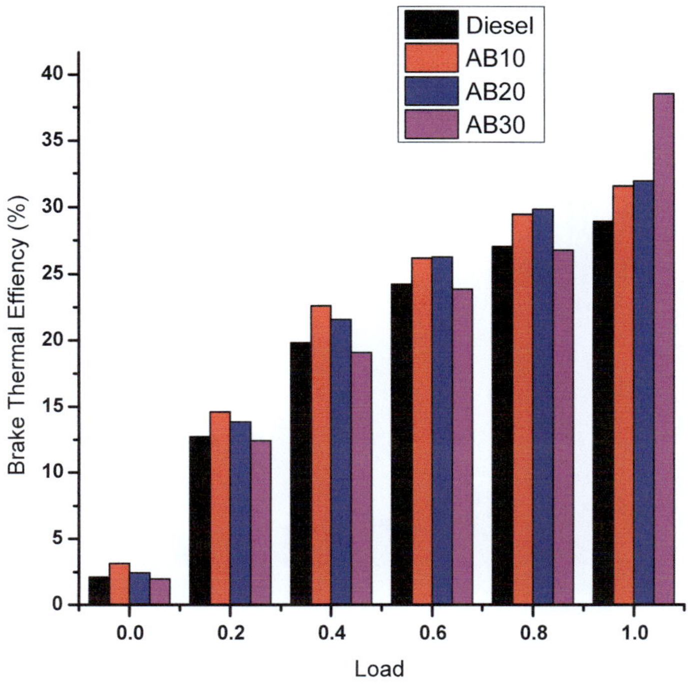

Fig. 3.13 Brake thermal efficiency versus load (%)

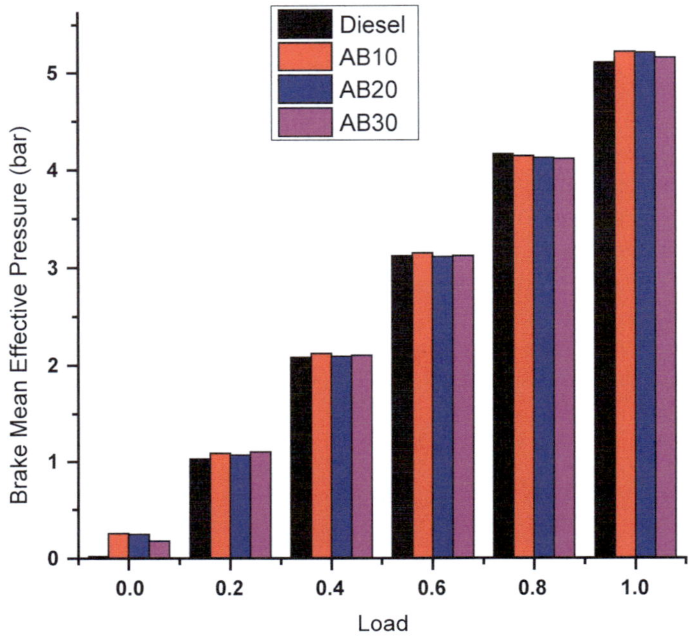

Fig. 3.14 Brake mean effective pressure versus load (%)

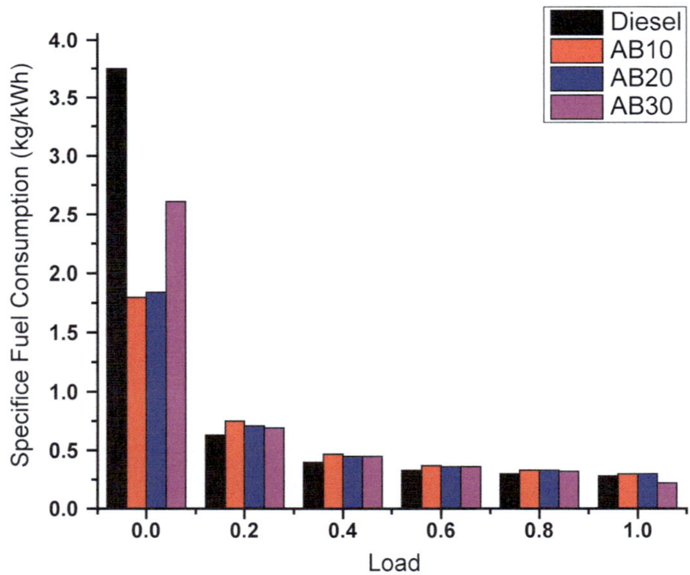

Fig. 3.15 Specific fuel consumption (kg/kWh) versus load (%)

conditions. At part load (20, 40, 60, 80%) conditions, specific fuel consumption was slightly increased than that of diesel fuel. The specific fuel consumption was minimum for AB30 at full load conditions than that of diesel fuel. This experiential fact is due to the excess amount of oxygen presents in the biodiesel and to give better combustion in the engine. The low calorific value of biodiesel leads to high specific fuel consumption rate than that of diesel fuel.

3.6.4 Air–Fuel Ratio (A/F Ratio)

The variation of A/F ratio versus different loading conditions was shown in Fig. 3.16. The air–fuel ratio was decreased with increase in load. At no load conditions, the A/F ratio was increased for diesel than that of all blends. The A/F ratio was least for AB10, AB20 than that of diesel fuel at full load conditions. The A/F ratio was maximum of 29.58 for AB30 at full load conditions. This is due to higher density and viscosity of fuel and low calorific value of the fuel.

3.6.5 Volumetric Efficiency

The variation of volumetric efficiency versus different load conditions was shown in Fig. 3.17. The volumetric efficiency was decreased due to increase in load on the

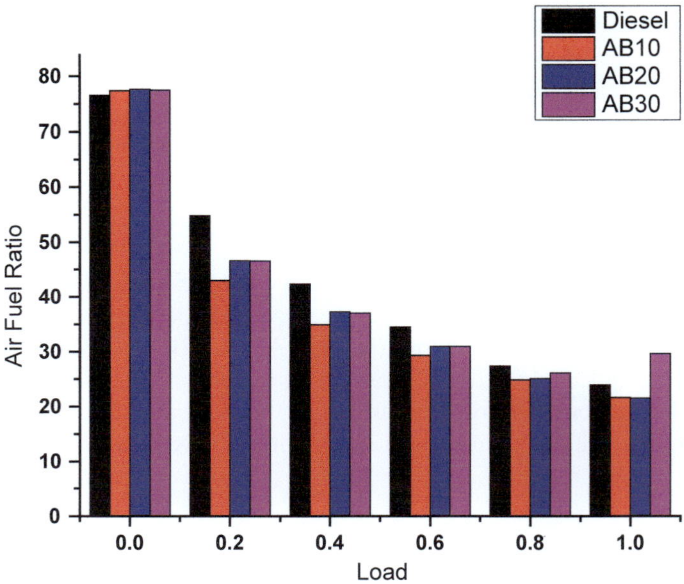

Fig. 3.16 A/F ratio versus load (%)

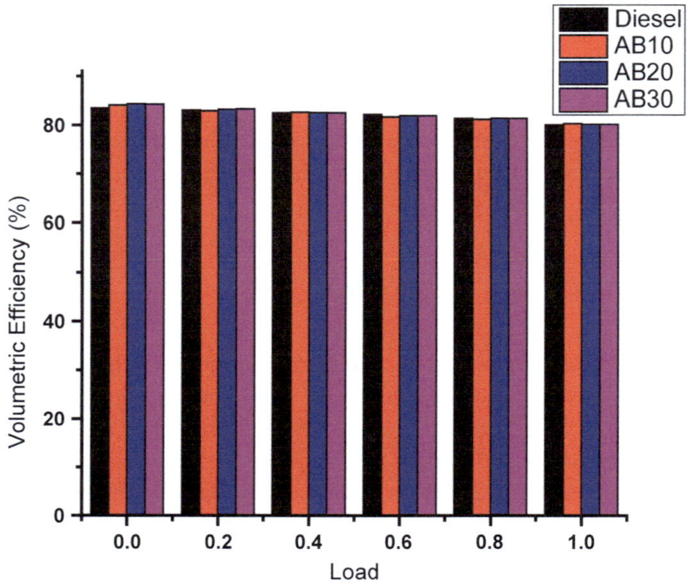

Fig. 3.17 Volumetric efficiency versus load (%)

engine. The volumetric efficiency was 83.54, 84.11, 84.31 and 84.26% for diesel, and AB10, AB20 and AB30 at no load conditions. But volumetric efficiency was decreased in increase in load. During part or full load conditions lower exhaust gas temperature released after the combustion process. At full load conditions, volumetric efficiency was 79.98, 80.27, 80.1, 80.1 for diesel and AB10, AB20 and AB30 than that of diesel fuel. The volumetric efficiency was slightly increase than that of diesel fuel at full load conditions.

3.6.6 Torque

The variation of torque versus different loading conditions was shown in Fig. 3.18. The torque of the engine with argemone biodiesel blends was increased with increased in different load conditions (no load, AB10, AB20, AB30). The values of torque were 26.86, 27.44, 27.36, 27.09 Nm for diesel and AB10, AB20 and AB30 at full load conditions. The torque of an engine was maximum for AB10 at full load conditions. The value of torque was increased due to lower viscosity and density of the fuel.

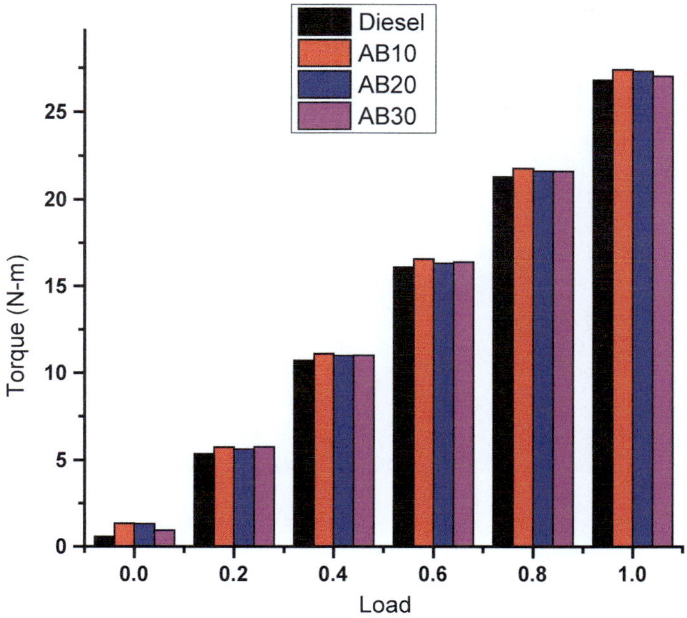

Fig. 3.18 Torque versus load (%)

3.6.7 Heat Loss Due to Jacket Water

The variation of heat loss by jacket water with different load conditions was shown in Fig. 3.19. The heat loss was decreased with increased in different load conditions. At part load conditions, heat loss by water jacket was increased due to increase load with increase BMEP and temperature of the water jacket increased. At full load conditions, heat loss was increased for AB30 compared than that of diesel and AB10, AB20.

3.6.8 Heat Loss by Gas

The variation of heat loss by gas with different load conditions was shown in Fig. 3.20. The heat loss was decreased with increased in load conditions at no load conditions. At part load conditions, heat loss by gas was increased due to increased load with increase due to higher temperature inside the cylinder and extra fuel is burnt to get together higher load command. At full load conditions, heat loss was increased for AB30 compared than that of diesel and AB10, AB20.

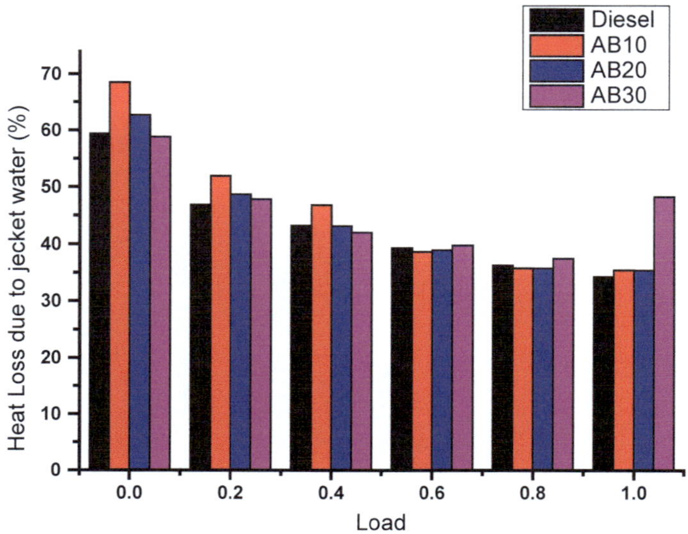

Fig. 3.19 Heat loss by jacket water versus load (%)

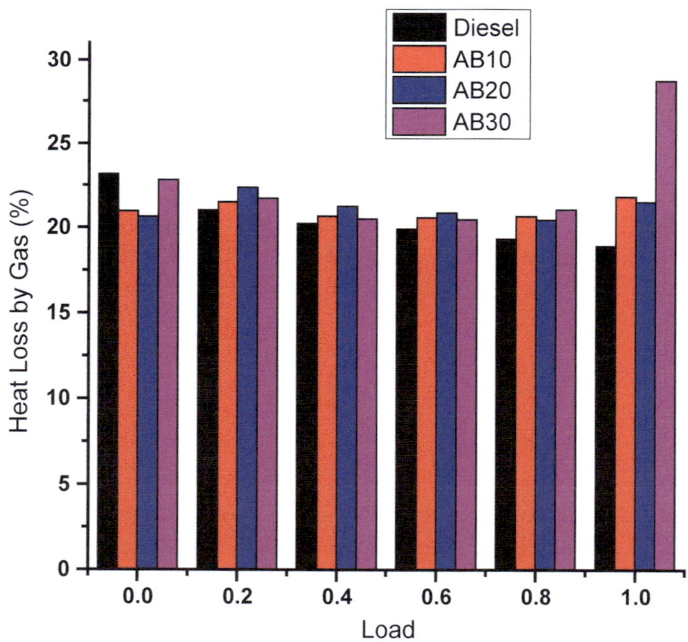

Fig. 3.20 Heat loss by gas versus load (%)

3.7 Emission Parameters

The exhaust gas emission with argemone biodiesel blends AB10, AB20 and AB30 had been evaluated in terms of CO_2, CO, HC, NO_x at different loading conditions (no load, part load and full load) of the engine.

3.7.1 Carbon Monoxides Emission

The variation of CO emission with respect to different load conditions was shown in Fig. 3.21. The CO emission was decreased with increased in load. The CO emission was maximum of 0.01 (vol%) for diesel at full load conditions than that of biodiesel blends. The CO emission was 100% reduced for AB10, AB20 and AB30 at part load and full load conditions. At part or full (60, 80, 100%) load conditions, CO emission was negligible, but in case of diesel fuel the value of CO emission was maximum. The CO emission was decreased at full load conditions due to the presence of oxygen content in the biodiesel. The oxygen content leads to better combustion and complete oxidation of fuel.

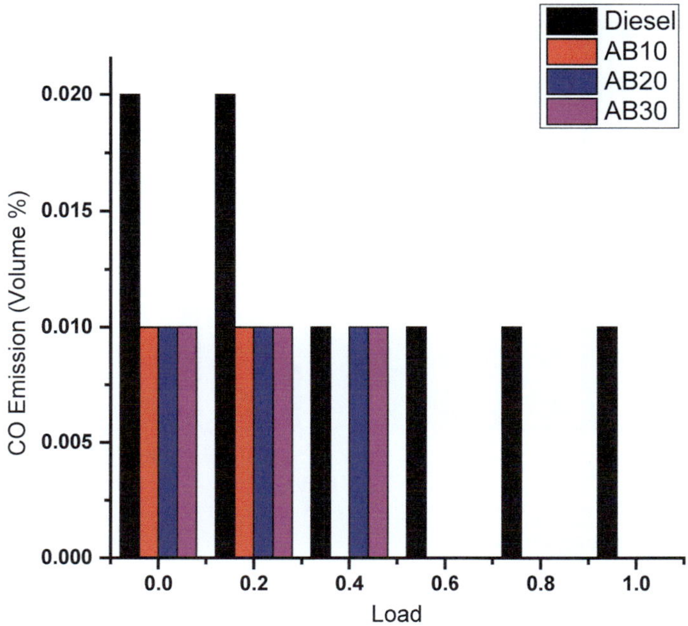

Fig. 3.21 CO (vol%) versus load (%)

3.7.2 Hydro Carbon Emission

The variation of HC emission with different load conditions was shown in Fig. 3.22. The HC emission was decreased with increased in load. The HC emission was increased at no load conditions but increasing in load on engine the value of HC was decreased. The values of HC were 6, 1, 1, 1 ppm for diesel AB10, AB20, AB30 at full load conditions and values of HC were 15, 3, 6 and 6 ppm for diesel, AB10, AB20 and AB30 at no load conditions. HC emission was decreased: 60, 66, 83 and 83% for diesel, AB10, AB20 and AB30 at full load conditions. The value HC emission was lower than that of mineral diesel. The HC value was decreased due to oxygen content present in the biodiesel. The oxygen content present in the fuel leads to better operation of fuel and complete combustion of fuel. The HC was reduced due to increased amount of biodiesel in the fuel blends because of biodiesel inherent oxygen content in the molecular arrangement for the whole combustion.

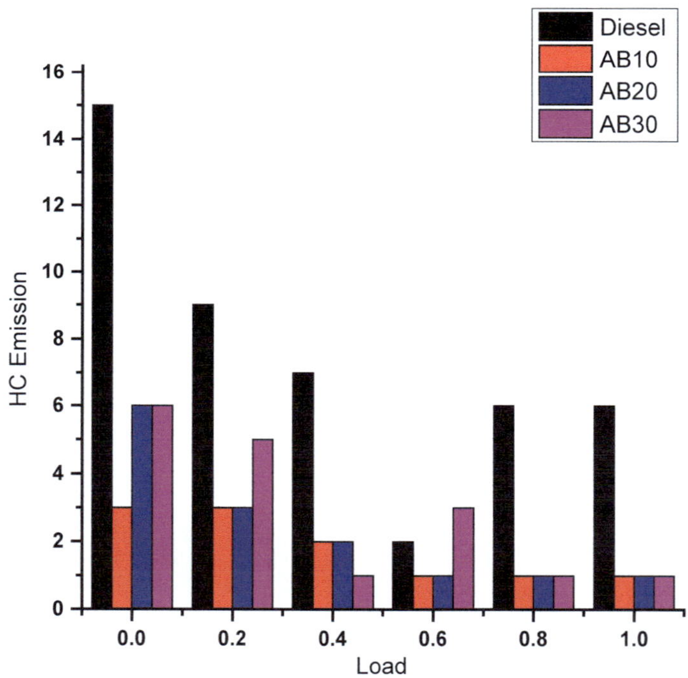

Fig. 3.22 HC (vol ppm) versus load (%)

3.7.3 Carbon Dioxide Emission

The variation of CO_2 emission with different load conditions was shown in Fig. 3.23. The CO_2 emission was decreased with increase in load. The CO_2 emission was reduced for increasing the load on engine. The values of CO_2 emission were 1, 0.5, 0.7 and 0.7% for AB0, AB10, AB10, AB20 and AB30 at full load conditions and also reduced at part load conditions than that of diesel fuel value. The value of CO_2 emission 0.5 (vol%) was highly decreased for argemone biodiesel blends AB10 at full load (100%) conditions. The CO_2 emission was decreased due to complete combustion and good spray atomization and uniform mixture formation compared than that of diesel.

3.7.4 Nitric Oxide

The variation of HC emission with different load conditions was shown in Fig. 3.24. The NO_x was increased with increased in load. The values of NO_x were 22, 15, 15 and 17 ppm for diesel, AB10, AB20 and AB30 at no load conditions. At 20% load, the values of NO_x were 41, 56, 50 and 53 ppm for diesel, AB10, AB20 and AB30. At part load conditions, the value of NO_x formation was slightly increased than that of diesel fuel. At full load conditions (100%), the values of NO_x

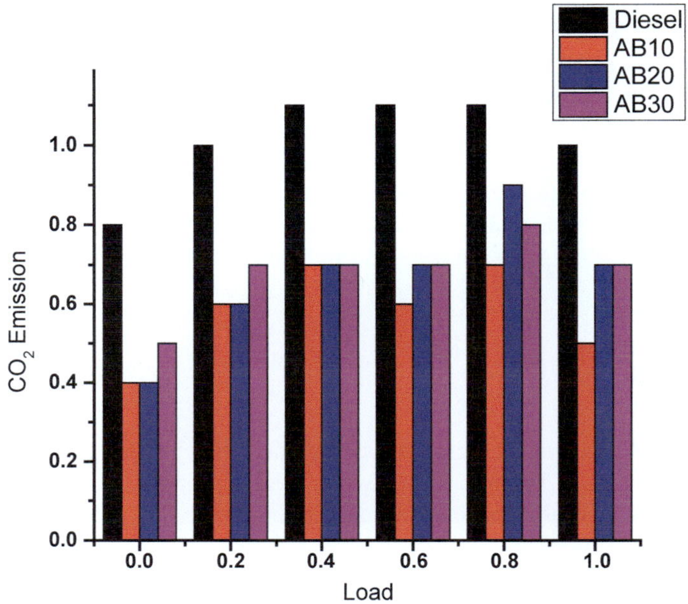

Fig. 3.23 CO_2 (vol ppm) versus load (%)

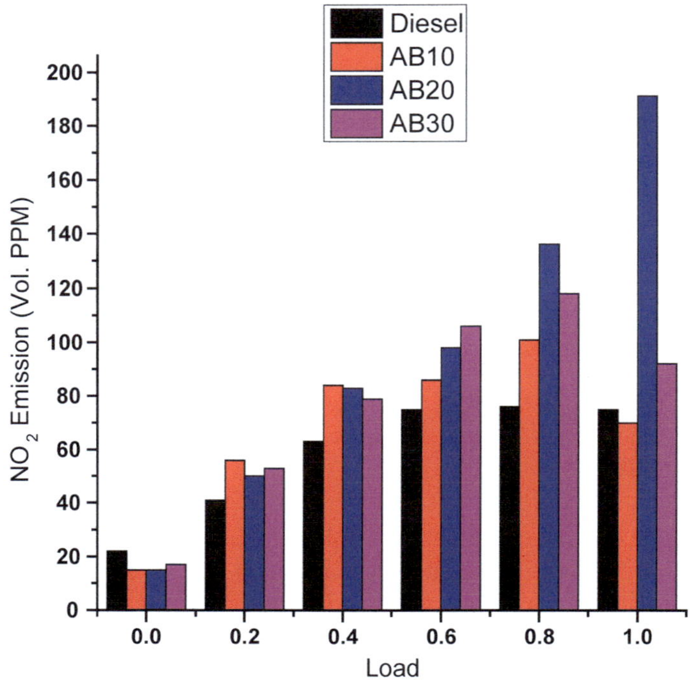

Fig. 3.24 NO_2 versus load (%)

emission were 75, 70, 191 and 92 ppm for diesel, AB10, AB20, AB30. The full load conditions of NO_x value were decreased for AB10 than that of diesel, AB20 and AB30. The biodiesel was provided extra oxygen content to the formation of NO_x emission. The NO_x emission was increased due to the excess amount of oxygen content present in the argemone biodiesel fuel. The NO_x emission was dependent on combustion temperature in the cylinder. The combustion temperature was increased due to increase the load on the engine. The excess oxygen content was leads to insufficient combustion and increased the NO_x formation.

3.8 Conclusion

On the basis of experimental result of performance and emission analysis on single cylinder, four-stroke CI engine fuelled with argemone biodiesel blends with diesel, the following conclusions have been made in this work:

- The most favorable condition for transesterification with methanol in the presence of homogeneous base catalyst sodium metal was molar ratio 1:6 (oil to methanol), catalyst concentration, reaction temperature 75 °C, reaction time 120 min, stirring speed 550 rpm.

- The physical and chemical properties of argemone biodiesel and blends with diesel were similar than that of diesel fuel.
- The brake thermal efficiency was 38.55% for AB30 at full load conditions. The brake thermal efficiency was increased for all blends AB10, AB20 and AB30 at different load conditions. This is due to 10–12% oxygen content present more than in diesel fuel.
- The indicated thermal efficiency was higher for AB10, AB20 and AB30 than that of diesel fuel at all load conditions. Higher indicated thermal efficiency was recorded 55.35% for AB30 at full load conditions.
- Higher the specific fuel consumption was recorded for AB10, AB20 and AB30 at part load conditions than that of diesel. The specific fuel consumption was decreased for AB30 at full load conditions compared than that of diesel due to viscosity, and density of biodiesel was similar to biodiesel blends.
- The A/F ratio was higher for AB30 at full load conditions than that of diesel fuel. The A/F ratio was lower for AB10 AB20 than that of diesel fuel for full load conditions. The A/F ratio for AB30 was increased due to higher viscosity and density of fuel.
- The heat released rate is increased for full load conditions because of the excess amount of oxygen content present in the biodiesel and due to its low calorific value, high density and high viscosity of fuel.
- Based on the emission analysis CO, CO_2, HC was decreased with increased for AB10, AB20, and AB30 at all load conditions the reason behind it oxygen content which contribute more complete combustion and reduced emission. The value CO, CO_2 and HC was lower than that of diesel.
- The NO_x emission was increased with increase in load. The NO_x emission was lower for AB10 at full load conditions than that of diesel fuel.
- High amount of oxygen content leads to more NO_x, and oxygen emission was produced.

References

1. Singh RK, Rath S (2011) Performance analysis of blends of karanja methyl ester in a compression ignition engine. IPCBEE 11:187–192
2. Nagarhalli MV, Nandekkar VM, Mohit KC (2010) Emission and performance characteristics of karanja biodiesel and its blend in CI engine and its economic. ARPN J Eng Appl Sci 5:52–56
3. Choudury S, Bose PK (2008) Karanja oil-its potential and suitability as biodiesel, 1–16
4. Rajan K, Senthilkumar KR (2009) The effect of exhaust gas recirculation on the performance and emission characteristics of diesel engine with sunflower oil methyl ester. Jordan J Mech Ind Eng 1:31–39

5. Siddalingappa R. Hotti. And Omparkash Hebbal. (2011) Performance and combustion characterstics of single cylinder diesel engine running on karaja oil/diesel fuel blends engineering, vol 3, pp 371–375
6. Agarwal AK, Rajamanoharan K (2009) Experimental investigations of performance and emissions of karanja oil and its blends in a single cylinder agricultural diesel engine. Appl Energy 86:106–112
7. Aydin H, Ilkılıc C (2010) Effect of ethanol blending with biodiesel on engine performance and exhaust emissions in a CI engine. Appl Therm Eng 30:1199–1204
8. Duran SK, Singh M, Singh H (2015) Karanja and rapeseed biodiesel: an experimental investigation of performance and combustion measurement for diesel engine. Int J Sci Eng Res 6:295–299
9. Devan PK, Mahalakshmi NV (2009) A study of the performance, emission and combustion characteristics of a compression ignition engine using methyl ester of paradise oil–eucalyptus oil blend. Appl Energy 86:675–680
10. Sonune PP, Farkade HS (2012) Performance and emission of CI engine fulled with preheated vegetable oil and its blends—a review. Int J Eng Innovative Technol 2:123–128
11. Sigar CP (2006) Comparative investigation on combustion and performance of a dual fuel engine using LPG and biodiesel (Karanja Oil). Ph.D. thesis, MNIT, Jaipur, India
12. Ramadhas AS, Jayaraj S, Muraleedharan C (2004) Use of vegetable oils as IC engine fuels—a review. Renew Energy 29:727–742
13. Vivak, Gupta AK (2014) Biodiesel production from karanja oil. J Sci Ind Res 63:39–47
14. Thirupathi Reddy K, Reddy PR, Ramana PV (2014) Effect of injection opening pressure on the engine performance with free fatty acid of rice bran oil as biodiesel. Int J Eng Res 3:179–182
15. Hossian MA, Chowdhury SM, Rekhu Y, Faraz KS, Islam MU (2012) Biodiesel from coconut oil: a renewable alternative fuel for diesel engine. World Acad Sci Eng Technol 6:1063–1067
16. Satyanarayana C, Rao PV (2009) Influence of key properties of Pongamia biodiesel on performance combustion and emission characteristics of a DI diesel engine. Wseas Trans Heat Mass Transf 4(2):34–44
17. Agarwal AK, Bijwe J, Das LM (2003) Effect of biodiesel utilization on wear of vital parts in compression ignition engine. J Eng Gas Turb Power 125:604–611
18. Pandey RK, Rehman A, Dixit S, Sarviya RM (2009) Performance and emission evaluation of diesel engine fueled with karanja oil. Int J Environ Res 3(3):463–470
19. Soni SL, Sharma SK, Sharma D, Khatri KK (2009) Optimization of injection timing and pressure of stationary CI engine operated on pre-heated karanja-diesel blend. Indian J Air Pollut Control IX:79–89
20. De Almeida SC, Belchior CR, Nascimento MV, Vieira LSR, Fleury G (2002) Performance of a diesel generator fuelled with palm oil. Fuel 81(16):2097–102
21. Rao RY, Zubaidha PK, Kondhare DD, Reddy NJ, Deshmukh SS (2012) Biodiesel production from argemone mexicana seed oil using crystalline manganese carbonate. Pol J Chem Technol 14:65–70

Chapter 4
Studies on Properties of Biodiesel Prepared from Microalgae (*Chlorella* sp.)

S.V. Kelaiya and P.M. Chauhan

Abstract Microalgae (*chlorella*) are an organism capable of photosynthesis and has less than 2 mm diameter in size. The biodiesel prepared from microalgae using chloroform–methanol extraction solvent system followed by three different trans-esterification processes (based on three different catalysts viz. Alkali catalyst, Acid catalyst, and Enzymatic catalyst) with two temperature (50 and 60 °C) and with 1:5 methanol to oil ratio. After transesterification, the fuel properties were measured and compared with the standard value of ASTM D 6751 standards. Alkali catalyst (0.56% NaOH) yields the highest production of biodiesel (92%) at 60 °C temperature. Also, the closest value of different fuel properties found at par with standard value of ASTM D 6751 standards viz. moisture content, carbon residue, calorific value, specific gravity, acid value, flash point, viscosity, density, and viscosity was found to be 0.01%, 0.04%, 40.41 MJ/kg, 0.83, 0.23 mg KOH/g, 143.67 °C, 5.16 mm^2/s, and 0.83 g/cm^3, respectively, in biodiesel.

Keywords Microalgae · *Chlorella sp.* · Biodiesel · Transesterification
Fuel properties

4.1 Introduction

Algae (macro and micro) have been taken in consideration as a residual biomass ready to be used for energy purposes. Algae, especially microalgae, were found to be the only source of renewable biodiesel that is capable of meeting the global demand for transport fuels [1]. The idea of using algae as a source of fuel is now being taken seriously because of the increasing price of petroleum and more

S.V. Kelaiya (✉) · P.M. Chauhan
Department of Renewable Energy Engineering, C.A.E.T., J.A.U, Junagadh, Gujarat, India
e-mail: sagarkelaiya@rediffmail.com

S. Kumar et al. (eds.), *Conference Proceedings of the Second International Conference on Recent Advances in Bioenergy Research*, Springer Proceedings in Energy, https://doi.org/10.1007/978-981-10-6107-3_4

significantly, the emerging concern about global warming that is associated with burning fossil fuels [4].

This work aimed to process development for production of biodiesel from microalgae, using two different extraction solvent systems to get the best result in oil extraction. Biodiesel is produced by trans-esterifying the parent oil or fat with an alcohol, usually methanol, in presence of a catalyst, usually a strong base such as sodium or potassium hydroxide or increasingly, alkoxides, acids, and enzymes. The resulting product therefore can contain not only the desired alkyl ester product but also unreached starting material (TAG), residual alcohol, and residual catalyst. Glycerol is formed as a by-product and separated from biodiesel in the production process; however, traces thereof can be found in the final biodiesel product. Since transesterification is a stepwise process, MAG and DAG formed as intermediates can also be found in biodiesel [6].

With the increasing interest and use, the assurance of fuel properties and quality has become of paramount interest to the successful commercialization and market acceptance of biodiesel. Accordingly, biodiesel standards have been established or are being developed in various countries and regions around the world, including the USA [2], Europe (EN 14214), Brazil, South Africa, Australia, and elsewhere [6]. In ASTM D 6751, and EN 14214 is commonly used standards as reference or base for other standards and their analysis.

4.2 Materials and Methods

Microalgae (*chlorella* sp.) were cultivated from locally available sources for the present study. The investigation methodology includes procedure for determination of properties of microalgae biodiesel and analysis of resulting data.

4.2.1 Experimental Design

The experiments were planned using completely randomized design (C.R.D.) [5]. The treatments consisted three levels of catalyst for transesterification, two levels of temperature, and two levels of solvent oil extraction methods in which best suitable method was adopted for oil extraction. The details of treatments and parameters are given as below.

- Year/Season of experiment: 2013–2014
- Crop: Algae, *Chlorella sp.*

4.2.2 Statistical Design: C.R.D.

- **First factor: Catalyst for transesterification**
  - (a) Alkali catalyst (C_1): Sodium Hydroxide (NaOH)
  - (b) Acid catalyst (C_2): Sulphuric Acid (H_2SO_4)
  - (c) Biocatalyst (C_3): Lipase Crude
- **Second factor: Temperature for transesterification**
  - (a) 50 °C (T_1)
  - (b) 60 °C (T_2)
- **Third factor: Solvents for oil extraction**
  - (a) Chloroform/Methanol (S_1)
  - (b) Hexane/Ether (S_2)

Total treatment combination: six and total no of observation (properties) is eight from each treatment with three replication.

4.2.3 Dependent Variable

(a) Quantity of biodiesel
(b) Quality of biodiesel

The results regarding the effect of solvent methods on oil recovery, effect of catalyst and temperature on biodiesel recovery were analyzed statistically [5].

4.2.4 Extraction Method of Algal Oil

In Chloroform–Methanol (2:1, v/v) Method, 10 g of dry algae was taken and mixed with chloroform–methanol (100 mL, 2:1, v/v) for 20 min using shaker followed by the addition of mixture of chloroform/water (50 mL, 1:1, v/v) for 10 min then it was filtered. While in hexane–ether (1:1, v/v) method, 10 g of dry algae was mixed with hexane–ether (100 mL, 1:1, v/v), keep it to settle for 24 h [3] and then it was filtered.

4.2.5 Transesterification and Biodiesel Production

In this method, oil to methanol ratio was 1:5, and the amount of all three catalysts was 1% by weight with oil. The reactions were started at two different temperatures likely 50 and 60 °C. After reaction took place, biodiesel and glycerin were separated by gravitational method using separating funnel. The produced biodiesel was tested for the different property measurement which is given below.

4.2.6 Analysis of Biodiesel

The fuel properties of algal biodiesel were determined by the following methods.

4.2.6.1 Moisture Content %

As per ASTM D2709, the moisture content of extracted biodiesel was determined by calculating the loss in weight of sample using hot air oven drying method.

$$\text{Moisture content}\,(\%) = \frac{W1 - W2}{W2} \times 100$$

where,

$W1$ Initial weight of biodiesel
$W2$ Final weight of biodiesel after drying

4.2.6.2 Carbon Residue %

As per ASTM D4530, the percentage of carbon residue content was determined. A 10 g of biodiesel was placed in a glass vial and heated to 500 °C under an inert (nitrogen) atmosphere in a controlled manner for 1 h. The mass% carbon residue in the original sample or in the 10% distillation bottoms was computed as follows:

Calculation for percent residue as follows:

$$\%\ \text{carbon residue} = \frac{(A \times 100)}{W}$$

where,

A carbon residue, g, and
W sample used, g

4.2.6.3 Calorific Value

The calorific value of biodiesel was determined using Bomb calorimeter (ASTM D240). Water equivalent (W) of calorimeter was obtained by conducting the experiment and using the benzoic acid as standard sample having known calorific value of 'H' equal to 6319 cal/g. The value of W was calculated by using the following formula.

$$H = \frac{\left(\text{wt of water, mL or g or } \frac{\text{cal}}{{}^\circ\text{C}}\right) + \left(W, \frac{\text{cal}}{{}^\circ\text{C}}\right) \times \text{temperature, } {}^\circ\text{C}}{\text{weight of sample, g}} - (E_1 + E_2)$$

Then, the calorific value of biomass will be obtained by conducting the separate experiment by using the following formula and the value of water equivalent, W, obtained from the above formula.

$$\text{Calorific value} = \frac{\left(\text{wt of water, mL or g or } \frac{\text{cal}}{{}^\circ\text{C}}\right) + \left(W, \frac{\text{cal}}{{}^\circ\text{C}}\right) * \text{temp.,}^{\circ\circ}\text{C}}{\text{weight of sample, g}} - (E_1 + E_2)$$

where,

$E1$ Correction for nichrome wire, cal = wt of wire, g × calorific value, 335 cal/g
$E2$ Correction for cotton thread, cal = wt of thread, g × calorific value 4180 cal/g

4.2.6.4 Specific Gravity

As per ASTM D4530, the specific gravity of biodiesel was determined by the following equation.

$$\text{Specific Gravity at } 30\ {}^\circ\text{C} = \frac{A - B}{C - B}$$

where,

A weight in gm of specific gravity bottle with oil at 30 °C
B weight in gm of specific gravity bottle at 30 °C
C weight in gm of specific gravity bottle with water at 30 °C

4.2.6.5 Acid Value

As per ASTM D664, acid value of biodiesel was determined using the following equation.

$$\text{Acid value} = \frac{56.1 \times V \times N}{W}$$

where,

V Volume in mL of standard potassium/Sodium hydroxide
N Normality of the potassium/Sodium hydroxide solution
W Weight in g of the sample

4.2.6.6 Flash Point

As per ASTM D93, the flash point of biodiesel was determined by Pensky Marten (Closed Cup) method.

4.2.6.7 Viscosity Measurement of a Liquid

As per ASTM D445, the viscosity of biodiesel was determined. The liquids of known densities were allowed to flow through the capillary maintaining the same differences of levels in the limbs and the time equation which governs the flow leads to the relation:

$$\frac{\eta_1}{\eta_2} = \frac{t_1 d_1}{t_2 d_2}$$

where, η_1 and η_2 are viscosity coefficients of the liquid and water, respectively. d_1 and d_2 are the densities of liquid and water, respectively. Knowing the value of viscosity of one liquid, one can calculate the viscosity of other liquid.

4.2.6.8 Density at 15 °C

As per ASTM D4052, the specific gravity of biodiesel was determined using the following equation.

$$\text{Density of biodiesel} = \frac{W}{V}$$

where,

W weight in gm of biodiesel at 15 °C
V volume in mL of pycnometer bottle at 15 °C

4.3 Results and Discussion

4.3.1 Production of Microalgae

Initially in open condition, 20 L of water was taken into 200 L capacity tank in which previously prepared culture (4 L), and media was added. Whole system was kept for 25 days for the growth of chlorella algae. The rate of production algae was found to be 1.5 g/L/day. Production data are presented in Table 4.1 (Figs. 4.1 and 4.2).

Table 4.1 Production practice of microalgae

Number of days	Production volume of culture (L)	Amount of media (g)	Production of wet biomass (g)	Yield of microalgae (dried powder) (g)
0–25	20	12.5	157	36
26–40	60	25	718	158
41–55	80	37.5	1421	341
56–70	100	50	2020	505
71–85	120	62.5	2741	603
86–100	140	75	3091	711
101–115	160	87.5	3317	796
116–130	180	100	4114	905
131–145	200	112.5	3879	1125
146–160	220	125	5041	1210
Total productivity			26,499	6390

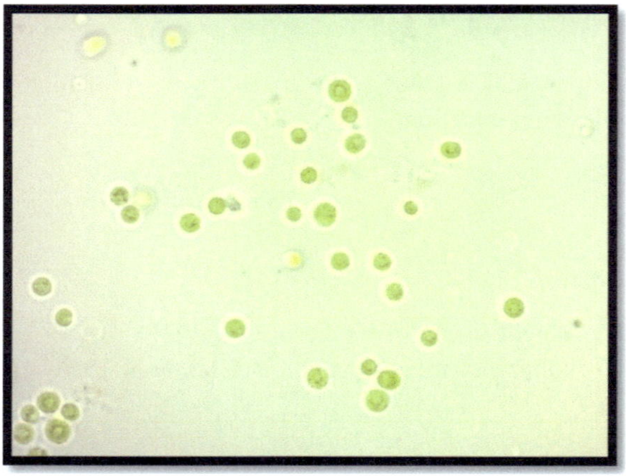

Fig. 4.1 Microscopic photograph of microalgae

Fig. 4.2 Pure microalgae culture (*Chlorella* sp.)

4.3.2 *Quality Analysis of Algal Biodiesel*

In the present study, two independent variables: type of catalyst and reaction temperature were evaluated considering three types of catalyst: Alkali catalyst, Acid catalyst, and Enzymatic catalyst; two levels of reaction temperature i.e. 50 and 60 °C. All the experiment combinations were carried out for three times in order to maintain the consistency of the results and to assess the experimental errors. The statistical data is given in Table 4.2 and the comparison of the properties of algal biodiesel prepared by different methods is given in Table 4.3 (Fig. 4.3).

4.3.3 *Analysis of Biodiesel*

4.3.3.1 Effect of Temperature (*T*) and Catalyst Type (*C*) on Different Properties of Biodiesel

An appraisal data on effect of catalyst type and temperature on different properties of biodiesel are presented in Table 4.3. Properties like moisture content, carbon residue, calorific value, specific gravity, acid value, flash point, viscosity, and density were found to be significant with change in catalyst type. Whereas significant properties with change in temperature were calorific value, specific gravity, acid value, flash point, viscosity, and density. In addition, the combined effect of the interaction between catalyst and temperature ($C \times T$) on carbon residue was found to be significant while calorific value, specific gravity, acid value, flash point, viscosity, and density were found to be highly significant while carbon residue and moisture content were found to be nonsignificant with change in temperature, and the combined effect of the interaction between catalyst and temperature ($C \times T$) on moisture content was found to be nonsignificant shown in Table 4.2.

The relationship between different biodiesel properties and different treatment combinations of biodiesel was shown in Fig. 4.4.

4.3.4 *Process Development for Algal Biodiesel Production*

The result presented in Table 4.3 and Fig. 4.4 shows that Alkali catalyst transesterification process is suitable for the production of biodiesel from microalgae

Table 4.2 Effect of reaction temperature (T) and catalyst (C) on fuel properties of biodiesel from microalgae

Property		Moisture content (%)	Carbon residue (%)	Calorific value (MJ/kg)	Specific gravity	Flash point (°C)	Acid value (mg KOH/g)	Viscosity (mm^2/s)	Density (g/cc)
Catalyst	Base (C_1)	0.01	0.04	40.11	0.86	137.00	0.27	5.68	0.86
	Acid (C_2)	0.03	0.06	41.09	0.90	114.83	0.41	7.04	0.90
	Enzyme (C_3)	0.03	0.02	40.31	0.86	142.50	0.22	5.66	0.86
	S.Em ±	0.00	0.00	0.11	0.00	0.67	0.01	0.01	0.00
	C.D. at 5%	0.01	0.01	0.35	0.01	2.05	0.02	0.04	0.01
Temperature	Temp. (T_1)	0.02	0.04	40.29	0.89	127.00	0.32	6.52	0.89
	Temp. (T_2)	0.02	0.04	40.72	0.86	135.59	0.28	5.73	0.86
	S.Em ±	0.00	0.00	0.09	0.00	0.54	0.01	0.01	0.00
	C.D. at 5%	NS	NS	0.29	0.01	1.68	0.02	0.03	0.01
$C \times T$	S.Em ±	0.00	0.00	0.16	0.01	0.94	0.01	0.02	0.01
	C.D. at 5%	NS	0.01	0.50	0.02	2.91	0.03	0.06	0.02
	C.V. %	23.14	10.07	0.69	1.15	1.24	5.42	0.55	1.15

Table 4.3 Property of biodiesel with different processes

Properties	Units	C_1T_1	C_1T_2	C_2T_1	C_2T_2	C_3T_1	C_3T_2	ASTM 6751	Methods and reference
Density	g/cc	0.89	0.83	0.91	0.9	0.84	0.87	0.86–0.9	D 4052
Viscosity	mm^2/s	6.2	5.16	7.18	6.9	5.14	6.17	1.9–6.0	D 445
Specific gravity	–	0.89	0.83	0.91	0.9	0.84	0.87		D 4052
Acid value	mg KOH/g	0.32	0.23	0.42	0.4	0.21	0.23	≤ 0.50	D 664
Flash point	°C	130.33	143.67	112.33	117.33	138.33	146.67	130.0 min	D 93
Moisture content	% (m/m)	0.01	0.01	0.03	0.02	0.03	0.04	≤ 0.05	D 2709
Carbon residue	% (m/m)	0.04	0.04	0.06	0.05	0.02	0.03	≤ 0.05	D 4530
Calorific value	MJ/kg	39.82	40.41	41.59	40.59	41.14	39.48	NR	D 240

NR Not reported in Specification by ASTM and ES or IS
C_1T_1: Alkali catalyst + Temperature 50 °C
C_1T_2: Alkali catalyst + Temperature 60 °C
C_2T_1: Acid catalyst + Temperature 50 °C
C_2T_2: Acid catalyst + Temperature 60 °C
C_3T_1: Enzymatic catalyst + Temperature 50 °C
C_3T_2: Enzymatic catalyst + Temperature 60 °C

C1T1 C2T1 C3T1 C1T2 C2T2 C3T2

Fig. 4.3 Biodiesel samples with different treatments

because of the highest biodiesel production and fuel properties of that treatment combination are match with ASTM standards.

The flow chart of the process developed for the production of biodiesel from microalgae is presented in Fig. 4.5.

4.4 Conclusions

The major conclusions were drawn during the process development of biodiesel from microalgae, as follows.

1. *Chlorlla* sp., as microalgae for the biodiesel production is used because it can be easily available in fresh water. It also has high oil percentage depends on cultivation practices.
2. The prepared biodiesel was analyzed for different fuel properties viz. moisture content, carbon residue, calorific value, specific gravity, acid value, flash point, viscosity, density, viscosity, and their values found to be 0.01%, 0.04%, 40.41 MJ/kg, 0.83, 0.23 mg KOH/g, 143.67 °C, 5.16 mm^2/s, and 0.83 g/cm^3, respectively. All the properties were found to be close to the standard value of ASTM D 6751 standards.
3. In the process development for the biodiesel production, the highest yield was obtained in the Alkali catalyst process which was 92% at the 60 °C temperature and 0.56% NaOH.

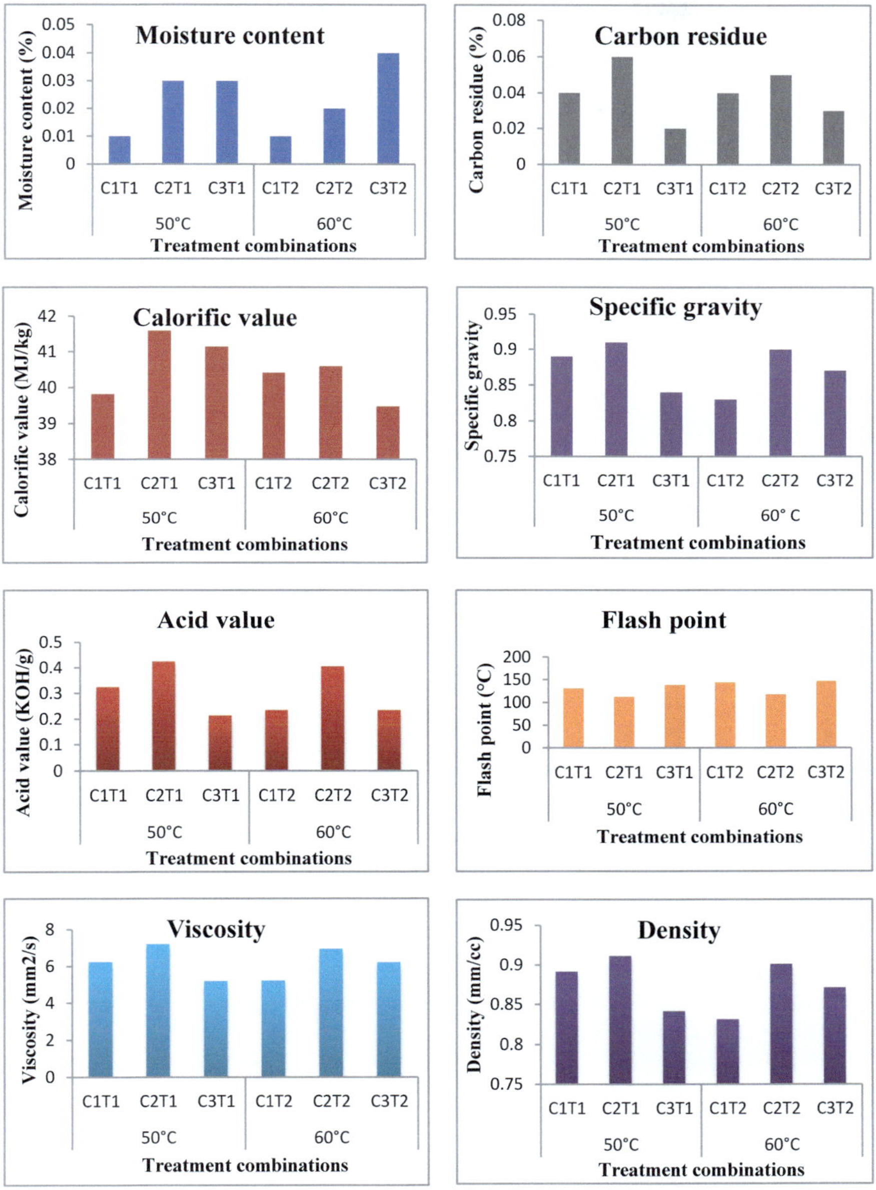

Fig. 4.4 Effect of temperature (*T*) and catalyst type (*C*) on different properties of microalgal biodiesel

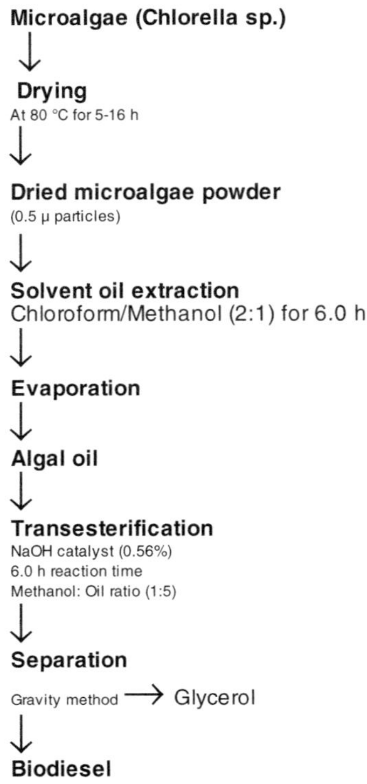

Fig. 4.5 Process flowchart for production of biodiesel from microalgae

References

1. Abd El-Moneim MR, Emad AS, Sanaa MS (2010) Enhancement of biodiesel production from different species of algae. grasas y aceites 61(4):416–422
2. ASTM D6751—American society for testing materials
3. Bligh EG, Dayer WJ (1959) A rapid method for total lipid extraction and purification. Can J Biochem Physiol 37:911–917
4. Gavrilescu M, Chisti Y (2005) Biotechnology a sustainable alternative for chemical industry. Biotechnol Adv 23:471–479
5. Gomez KA, Gomez AA (1984) Statistical procedures for agricultural research. Wiley, New York
6. Knothe G (2006) Analyzing biodiesel: standards and other methods. J Am Oil Chem Soc 83:823–833

Chapter 5
Ultrasound-Assisted Biodiesel Production Using KI-Impregnated Zinc Oxide (ZnO) as Heterogeneous Catalyst: A Mechanistic Approach

Ritesh S. Malani, Sohan Singh, Arun Goyal and Vijayanand S. Moholkar

Abstract As a consequence of fast reduction of fossil fuel reservoirs in addition to environment concerns of global warming, the demand for clean and renewable alternate fuel has increased in recent years. Biodiesel comes as assuring alternative liquid fuel produced by transesterification of triglycerides in presence of catalyst. In present work, a new solid catalyst of KI-impregnated ZnO has been developed using wet impregnation method. This solid catalyst has been tested for biodiesel synthesis using soybean oil and methanol. The catalyst prepared was characterized by X-ray diffraction (XRD), and the surface area of catalyst had been determined using Surface area and pore size analyzer. Transesterification using this catalyst has been carried out in presence of 35 kHz ultrasound. Response surface methodology (RSM) using central composite design (CCD) has been employed to assess the effect of three different process variables viz., methanol to oil molar ratio, catalyst wt% loading and reaction temperature. The optimized set of parameters have been determined as: catalyst loading = 6.15 wt%; methanol to oil molar ratio = 10.25:1 and temperature = 335.5 K, for which the highest conversion of 94.71% of oil was obtained. The catalyst retained 54.5% of the original (or fresh) after five cycles. Biodiesel synthesized with new catalyst was tested for fuel properties such as

R.S. Malani (✉) · A. Goyal · V.S. Moholkar
Centre for Energy, Indian Institute of Technology Guwahati,
Guwahati 781039, Assam, India
e-mail: r.malani@iitg.ernet.in

S. Singh
Department of Petroleum Engineering, D.I.T. University,
Dehradun 248009, Uttarakhand, India

A. Goyal
Department of Biosciences and Bioengineering, Indian Institute of Technology Guwahati,
Guwahati 781039, Assam, India

V.S. Moholkar
Department of Chemical Engineering, Indian Institute of Technology Guwahati,
Guwahati 781039, Assam, India

S. Kumar et al. (eds.), *Conference Proceedings of the Second International Conference on Recent Advances in Bioenergy Research*, Springer Proceedings in Energy, https://doi.org/10.1007/978-981-10-6107-3_5

calorific value, kinetic viscosity, specific gravity, flash, and fire point. These properties have been found to be within specific limits described by IS 15607:2005 and ASTM D6751 standards for commercial applications.

Keywords Ultrasound · Biodiesel · Heterogeneous catalyst · Potassium iodide Zinc oxide

5.1 Introduction

Energy is the basic need for the economic development of any country. The principal energy demands in all sectors of a developing country are in two forms, viz. electricity and transportation fuel [15]. Renewable energy has great potential to meet our future energy demands. Biodiesel is basically alkyl esters of fatty acids shown a great potential as alternative transportation fuel. These can be produced by esterification or transesterification of free fatty acid or triglycerides reacting with short-chain alcohols such as methanol and ethanol [3–5, 9]. The common feedstock used for esterification or transesterification reaction is oil derived from edible and non-edible source, waste cooking oil, animal fat etc. [10]. The common catalyst used for the esterification or transesterification reactions are alkali and acid [2, 3]. The homogeneous catalysts were more popular because of faster kinetics rate. On the other hand, these catalysts increase the cost of purification of biodiesel and also contaminate the by-product of glycerol [3, 4, 8, 12]. Heterogeneous catalysts offer a feasible solution to these problems, which can be separated without difficulty from the reaction mixture. However, slower kinetics and 3-phase (Catalyst (Solid)—Oil (Organic)—methanol (Aqueous)) heterogeneity of the reaction system reduces the biodiesel production rate [2–4, 8].

Many studies have been carried out using heterogeneous alkali catalysts such as alkali and alkaline earth oxides like CaO, $Ba(OH)_2$, basic zeolites [1, 11, 13, 19–21, 24]. Recent research has dedicated significant attempt to overcome slow kinetics of conventional heterogeneous catalytic processes. The activity of the heterogeneous catalyst can be increased by doping alkali earth ions to a suitable support like alumina and metal oxides [8, 23]. Mass transfer limitation in 3-phase reaction system (liquid–liquid–solid) also contributes to slow kinetics of transesterification with heterogeneous catalyst. Mass transfer limitation can be overcome by intense mixing, application of high temperature or pressure or use of co-solvent [14]. The application of ultrasound gives a possible solution to enhance the kinetics of mass transfer limited systems, as it creates intense micro-mixing in the medium [3, 4, 16].

In this paper, we have reported the development of a new heterogeneous catalyst using a wet impregnation method. We have characterized the new catalyst and applied it for biodiesel production. Limitations of the slower kinetic rate and limiting mass transfer overcome successfully by the application of ultrasound using conventional ultrasound bath. The experiments were designed using the statistical approach of central composite design (CCD) with Minitab 16 trial version, and responses have been used to optimize the reaction parameters [6, 7, 17]. The biodiesel produced using this catalyst was tested for different fuel properties. In addition, the reusability of the catalyst in successive cycles of transesterification was also studied.

5.2 Materials and Methods

5.2.1 Catalyst Preparation

Zinc oxide (AR, 98% Himedia, India) was used as a support material for the catalyst, and potassium iodide (AR, 99% Himedia, India) was selected as a source of potassium. The catalyst was made by wet impregnation method using 35% KI solution [22]. Twenty-five milligram of ZnO was added to 75 ml of 35% KI solution and stirred continuously for 3 h under constant heating. The semisolid impregnate was kept overnight in hot air oven at 383 K for drying, and further calcination was done at 773 K for 3 h in programed muffle furnace with temperature ramp of 10 °C/min [8, 22]. Subsequently, the catalyst was cooled to ambient temperature and was stored in vacuum desiccator. The catalyst was characterized by X-ray diffraction (XRD) for identification of different phases as well as the surface area of the catalyst was determined using surface area and pore size analyzer (Quantachrome, Model: Autosorb-IQ MP).

5.2.2 Transesterification Reactions

Fortune® edible soybean oil was bought from local market. Average molecular weight and acid value of soybean oil were found to be 880.56 g/mol and 0.1 mg KOH/g. Other chemicals used for transesterification were as follows: methanol (AR grade, 99% Merck India). The anhydrous methanol was obtained from vacuum distillation of methanol, which was used in all experiments.

Preliminary experiments: The as-prepared catalyst was tested for biodiesel synthesis using soybean oil with methanol. The reaction was brought in a 50-ml two-neck round bottom flask attached with reflux coil condenser using mechanical shaking (400 rpm) at 333 K. The total reaction volume was 25 ml with methanol to oil molar ratio of 10:1 and catalyst loading of 5 wt%. The reaction conducted for 1 h. At the end of 1 h of reaction, the mixture was filtered, and glycerol was separated using centrifuge, followed by water washing to remove unreacted methanol, and product was stored for analysis under refrigerated condition. The gross conversion of triglycerides to biodiesel was analyzed using ^{1}H nuclear magnetic resonance (NMR) technique (600 MHz Bruker) with $CDCl_3$ (Merck, India) as a solvent and TMS (tetramethylsilane) as internal standard [5]. After 1 h reaction, the conversion obtained in control experiment was very less, ~23%, which was essentially due to limited mass transfer limitations between the 3-phase of reaction system. In view of this result in preliminary experiments, the main experiments were conducted with application of ultrasound, as explained in greater details below.

5.2.3 Experimental Setup and Protocol

The schematic of experiment was shown below in Fig. 5.1. Transesterification experiments were carried out using an ultrasound bath (Make: Elma Trans-sonic T-460 type, Germany, Capacity: 2 L, frequency: 35 kHz, power: 35 W).

Millipore water was used as a medium transmission of ultrasound by filling the bath up to two-third of its total volume. Fifty-milliliter double-neck round bottom flask was used to carry out all the experiment. The round bottom flask was made of borosilicate glass fitted with a coiled condenser for refluxing of the methanol vapors. The reaction bath temperature was controlled at desired level using continuous water circulation at constant temperature through circular bath (Make: Jeio Tech, Model: Lab Companion RW 0525G). Position of reaction flask in the bath in all experiments was carefully maintained same, as the power (or pressure amplitude) of ultrasound waves shows momentous spatial deviation in the bath [18].

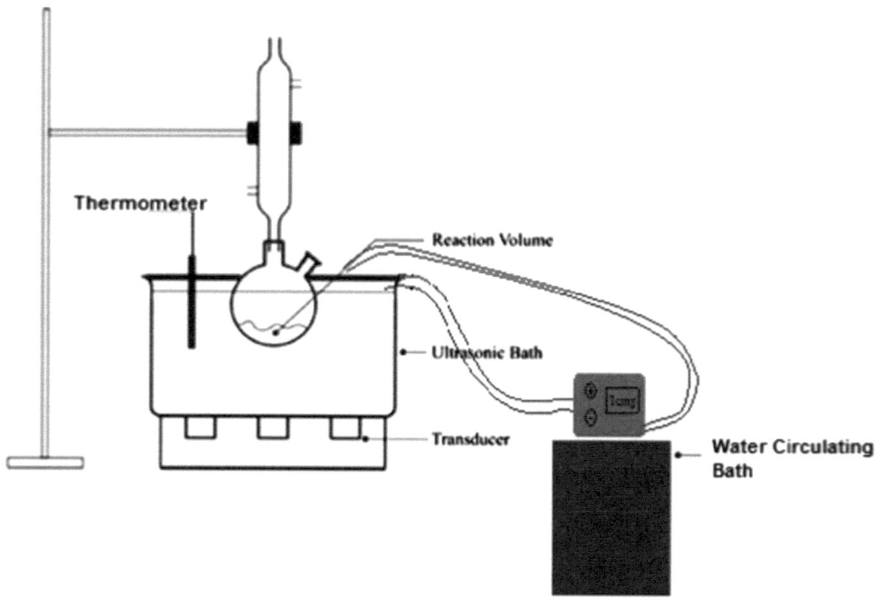

Fig. 5.1 Schematic of experimental setup

The transesterification experiments were planned using statistical central composite design (CCD) with three factors (or experimental parameters) at three levels. The CCD experimental design shown in Tables 5.1 and 5.2 included 20 sets with permutation/combination of the different levels of experimental parameters. This design was produced using software MINITAB 16 (Trial version). All the experiments were performed in triplicates, and the average values of triglyceride conversion have been mentioned in Table 5.2 with standard deviation. The analysis of the results CCD experimental design was done through response surface methodology (RSM) with a quadratic model. The advancement of transesterification reaction was supervised by withdrawal of 500 μl aliquots of reaction solution at every 10 min interval, followed by centrifugation for 5 min at 6000 rpm to distinct

Table 5.1 Range and levels of experimental process variables

Process variables (or factors)	Symbol (code)	Levels of factors coded value (actual value)		
Catalyst loading (wt% oil)	C	−1 (3)	0 (6)	+1 (9)
Temperature (K)	T	−1 (323)	0 (333)	+1 (343)
Alcohol:oil molar ratio (M)	M	−1 (5:1)	0 (10:1)	+1 (15:1)

Table 5.2 CCD statistical experimental sets

S. No.	Catalyst loading (wt% oil)	Molar ratio	Temperature (K)	% Triglyceride conversion (experimental)	Fitted % triglyceride conversion
1	6.00	5.00	333.00	54.40 ± 0.79	54.34
2	7.78	7.03	338.95	69.20 ± 1.11	68.47
3	6.00	10.00	333.00	91.81 ± 0.89	90.90
4	4.22	7.03	338.95	66.78 ± 1.33	67.76
5	6.00	10.00	333.00	90.40 ± 1.33	90.90
6	6.00	10.00	333.00	92.20 ± 1.37	90.90
7	3.00	10.00	333.00	74.78 ± 0.88	75.69
8	9.00	10.00	333.00	78.20 ± 0.44	77.71
9	6.00	10.00	323.00	16.80 ± 0.75	17.39
10	4.22	12.97	327.05	41.20 ± 0.69	41.64
11	6.00	10.00	333.00	89.98 ± 1.08	90.90
12	7.78	7.03	327.05	40.04 ± 1.15	41.28
13	6.00	15.00	333.00	59.80 ± 1.17	60.27
14	4.22	12.97	338.95	74.30 ± 2.04	72.76
15	6.00	10.00	333.00	90.28 ± 1.59	90.90
16	6.00	10.00	333.00	90.83 ± 1.42	90.90
17	7.78	12.97	338.95	74.70 ± 1.99	76.04
18	6.00	10.00	343.00	66.60 ± 1.32	66.43
19	7.78	12.97	327.05	44.60 ± 1.84	43.33
20	4.22	7.03	327.05	43.80 ± 1.41	42.16

the catalyst and glycerol form organic layer. The organic layer was washed with 1 ml of hot water followed by centrifugation at 6000 rpm for 5 min to remove unreacted methanol. The final organic or biodiesel layer was then stored at refrigerated conditions and further analyzed to determine the triglyceride conversion or biodiesel yield.

5.2.4 *Kinetics of the Reaction*

Stoichiometrically, in transesterification, 3 mol of methanol are required for conversion of 1 mol of triglyceride. However, in order to enhance conversion of triglycerides, excess alcohol is used. In the present study, we have employed large excess methanol for transesterification, as evident from the alcohol to oil molar ratios mentioned in Table 5.1. Due to large excess methanol, the overall concentration of methanol in reaction mixture shows insignificant variation. Therefore, the overall rate of transesterification reaction is essentially a function of concentration of only triglycerides. Hence, we have assumed pseudo first-order kinetics for the overall transesterification reaction, without taking into consideration for intermediate steps of conversion of triglycerides, and the below-given equation has been used to fit the conversion—time (X vs. t) data:

$$\ln(1 - X) = -kt$$

where X = conversion of triglyceride at time t. Plot of $-\ln(1 - X)$ versus t gives the kinetic constant k as the slope.

5.3 Result and Discussion

5.3.1 *Characterization of Catalyst*

XRD analysis: As mentioned earlier, the XRD pattern of native (or original) ZnO and ZnO impregnated with KI has been examined to identify different phases present in the catalyst. The X-ray spectra are shown in Fig. 5.2. The diffraction pattern of ZnO used in the present study matches with standard pattern with peaks at $2\theta = 32°$; $36°$; $48°$; $56°$; $64°$; $69°$; and $70°$ (JPDS file no. 36-1451). The peaks at $2\theta = 21°$; $25°$; and $35°$ show the presence of KI with high crystallinity [22]. In addition, peaks corresponding to K_2O phase are observed at $2\theta = 51°$ and $56°$. Literature has reported that the presence of K_2O phase on catalyst surface imparts basicity to the catalyst surface. The formation of Zn–O–K groups on catalyst surface due to K_2O phase serves as active sites for the transesterification reaction [22].

Pore size and surface area analysis: Original ZnO and KI-impregnated ZnO catalyst were analyzed for pore size distribution and surface area using nitrogen

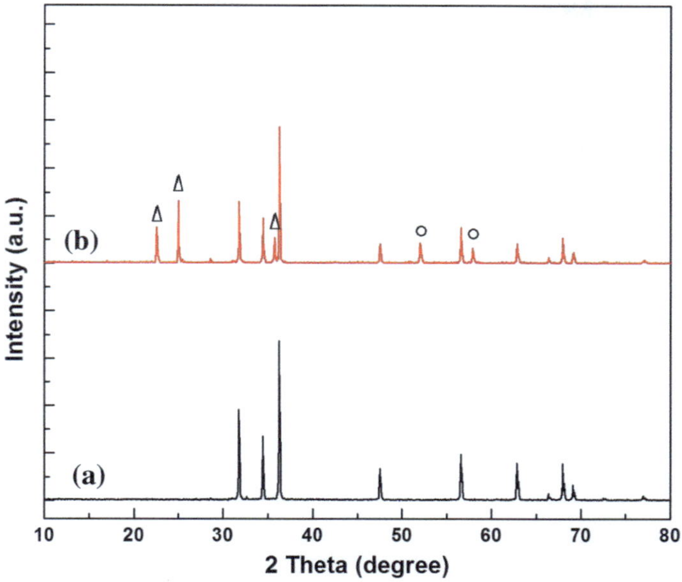

Fig. 5.2 XRD pattern of **a** ZnO and **b** KI/ZnO samples. (*Δ*) KI; (*o*) K_2O

adsorption and desorption isotherm. The average pore diameter of ZnO and KI/ZnO catalyst was 1.876 and 2.121 nm, respectively. The surface area analysis of ZnO and KI/ZnO yields the BET surface area as 16.667 and 3.853 m^2/g, respectively. The total pore volume of ZnO and KI/ZnO was determined as 0.022 and 0.008 cm^3/g, respectively.

5.3.2 Experimental Results and Statistical Optimization

The conversion of triglyceride to biodiesel was monitored with ^1H NMR spectra of the aliquots of reaction mixture. Figure 5.3 shows ^1H NMR spectrum of biodiesel from soybean oil. A peak at 3.6 ppm, as visible in Fig. 5.3, corresponds to FAME. Different peaks in the ^1H NMR spectrum have been characterized as follows: Triplet at 2.3 ppm agrees the presence of α-carbonyl methylene, whereas peaks associated to unsaturation in soybean oil are 2.0, 2.8, and 5.3 ppm, normally named to allylic, bis-allylic, and olefinic hydrogen, respectively. A strong singlet peak at 1.2 ppm denotes methylene group. The triplet arising at 0.8 ppm was associated with terminal methyl protons. The multiplet at 1.6 ppm was assigned to α-carbonyl methylenes. A peak at 2.72 ppm represents polyunsaturated fatty acids (methylene group between two double bonds of a higher polyunsaturated fatty acid chain). The gross yield of biodiesel from ^1H NMR spectrum has been calculated using following equation [5]:

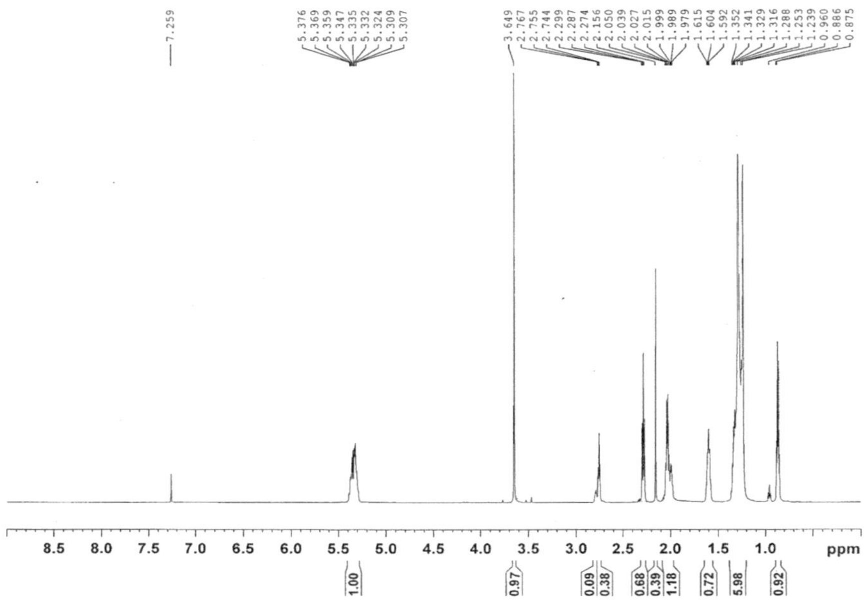

Fig. 5.3 ^{1}H NMR spectra of fatty acid methyl ester from soybean oil

$$X = (2 \times A_{ME}) \times 100/(3 \times A_{\alpha-CH_2})$$

where A_{ME} = integration value of strong singlet peak at 3.6 ppm, $A_{\alpha\text{-}CH2}$ = integration value of triplet at 2.3 ppm.

Optimization of transesterification parameters using RSM: The results of the CCD statistical experimental design are given Table 5.2 that depicts the experimental and predicted values of the response variable. The average triglyceride conversion in each experimental set has been given along with its standard deviation of the three runs. The coefficients obtained were fitted to quadratic regression model (depicted in Eq. below) using coded values for the process parameters (Minitab 16 software, trial version). The fitted model equation as follows:

$$Y = 90.9031 + 0.6013C + 1.7619M + 14.5783T - 5.0223C^2 - 11.8777M^2 \\ - 17.8225T^2 + 0.6425C \times M + 0.3975T \times C + 1.3825M \times T$$

Notation: T—reaction temperature, M—alcohol to oil molar ratio and C—catalyst concentration. The significance of every coefficient in the model equation has also been verified by standard t-test and p-values, also indicated in Table 5.3a. Values of triglyceride conversion forecasted by the quadratic model matches satisfactory with the results obtained from experiments. The high value (99.80%) of coefficient of determination (R^2) signifies the applicability of the selected model. Table 5.3b shows the ANOVA for the quadratic model.

Table 5.3 Statistical analysis of experiments

(a) Regression coefficients determined for triglyceride conversion

Term	Coefficients	SE coeff	*t*-stat	*p*-value
Constant (β)	90.9031	0.5373	169.194	0.000
Catalyst (C)	0.6013	0.3565	1.687	0.003
Molar ratio (M)	1.7619	0.3565	4.943	0.001
Temperature (T)	14.5783	0.3565	40.897	0.000
Catalyst × catalyst (C^2)	−5.0223	0.3470	−14.473	0.000
Molar ratio × molar ratio (M^2)	−11.8777	0.3470	−34.229	0.000
Temperature × temperature (T^2)	−17.3225	0.3470	−49.919	0.000
Molar ratio × catalyst (MC)	0.6425	0.4657	1.380	0.108
Temperature × catalyst (TC)	0.3975	0.4657	0.853	0.013
Temperature × molar ratio (TM)	1.3825	0.4657	2.968	0.014

(b) Analysis of variance (ANOVA) for triglyceride converion

Source	DF	Sq SS	Adj SS	Adj MS	F	*p*-value
Regression	9	8879.20	8879.20	986.58	568.52	0.000
Linear	3	2949.77	2949.77	983.26	566.60	0.000
Square	3	5909.58	5909.58	1969.86	1135.13	0.000
Interaction	3	19.86	19.86	6.62	3.81	0.000
Residual error	10	17.35	17.35	1.74	–	–
Lack-of-Fit	5	13.37	13.37	2.67	3.36	0.105
Pure error	5	3.98	3.98	0.80	–	–
Total	19	8896.55			–	–

R^2 = 99.80%; R^2 (adj) = 99.63%

Figure 5.4 represents the response surface 3-D plots which show the interaction effects between two variables holding the third variable at midpoint value.

5.3.3 *Kinetic Analysis of the Reaction*

As said earlier, the time profile of triglyceride conversion was fitted to pseudo first-order reaction kinetic equation. Figure 5.5 depicts experimental data and pseudo first-order model fit for the experiments carried out at alcohol to oil molar ratio = 10 and catalyst loading = 6 wt% at three altered temperatures. The values of the pseudo first-order kinetic constants along with regression coefficients are given in Table 5.4. The regression values for all kinetic constants at all temperatures are >0.9, which indicates that overall kinetics of the three-step transesterification process is well described by the pseudo first-order model based on time profile of the limiting reactant of triglyceride.

Reusability of catalyst: KI-impregnated-ZnO catalyst in the present study was also subjected to reusability test. The catalyst used for transesterification reaction in

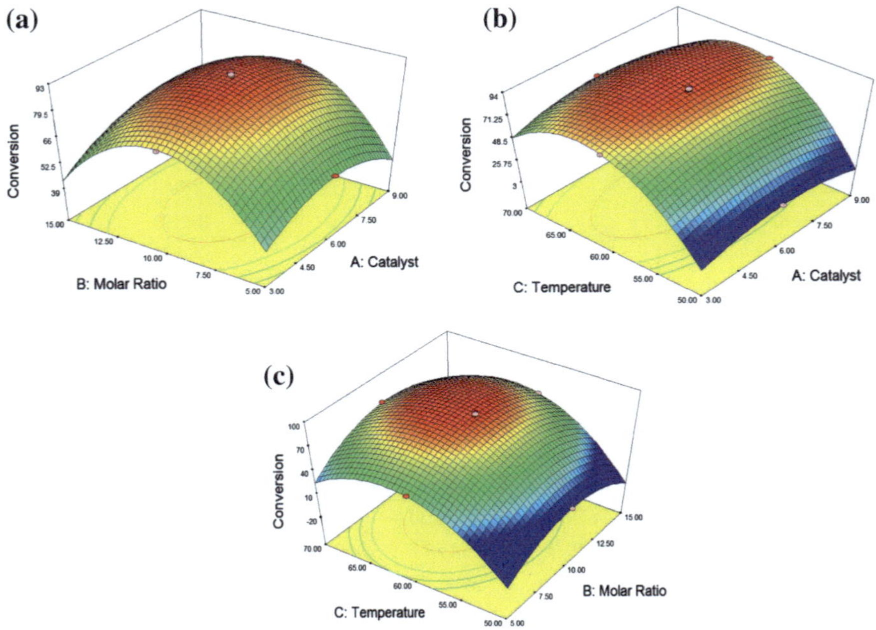

Fig. 5.4 3-D response surface plots for triglyceride conversion

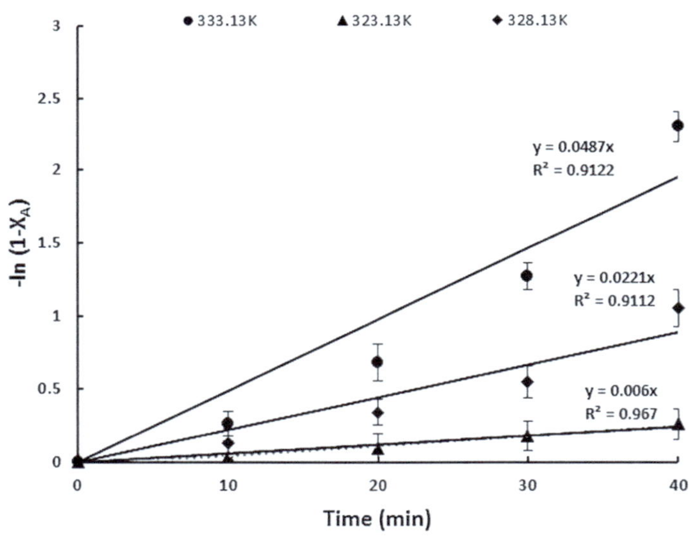

Fig. 5.5 Pseudo 1st order reaction kinetic model fitted to transesterifctaion reaction data

Table 5.4 Kinetic investigation of trasesterifcation reaction

Temperature (K)	Rate constant (k) (min^{-1})
323	0.0060 (R^2 = 0.967)
328	0.0221 (R^2 = 0.911)
333	0.0487 (R^2 = 0.912)

first cycle was recovered by centrifugation of the reaction mixture. This catalyst was subjected to washing with n-hexane, which removes the impurities from the surface such as methanol, oil, or glycerol. The washed catalyst was heated in hot air oven for 3 h at 383 K and was reused for transesterification reaction. This procedure was followed for five cycles, and the activity of catalyst was analyzed on the basis of triglyceride conversion as compared to initial conversion. Figure 5.6 shows the results of the reusability test of catalyst in the form of biodiesel yield in successive cycles with respect to fresh catalyst. Results depicted in Fig. 5.6 reveal that catalyst retained over 50% of its initial activity after five cycles of successive transesterification.

Properties of FAME: The biodiesel synthesized using KI/ZnO catalyst was subjected for fuel properties testing such as specific gravity, kinetic viscosity, flash, and fire point calorific value. Kinematic viscosity and density of biodiesel were lesser than soybean oil. Table 5.5 summarizes the properties of biodiesel samples obtained with 6.15 wt% KI/ZnO catalyst to molar ratio 10.25:1 (methanol to oil) and temperature 335.5 K. All the values presented in the Table 5.5 show results within the standards prescribed by IS 15607:2005 and ASTM D6751 standards.

5.3.4 Discussion

Statistical optimization helps us identify various facets of different independent operating parameters. As seen from the ANOVA results, F-values of linear

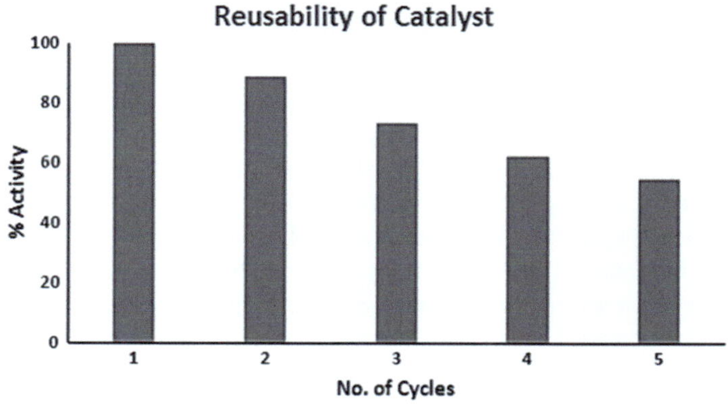

Fig. 5.6 Reusability results of catalyst for transesertication reaction

Table 5.5 Physico-chemical analysis biodiesel fuel

Analysis	Method	Biodiesel	Specification
Specific gravity (kg/m^3)	ASTM D 4052	887.4	850–900
Kinetic viscosity (mm^2/S^2)	ASTM D 445	4.88	3.0–6.0
Flash point (°C)	ASTM D 93	185	>120
Fire point (°C)	–	203	–
Calorific value (kJ/kg)	–	28722	–

coefficients (indicating the significance of individual parameters) are far higher than the F-values of interaction parameters. This essentially points that the three optimization variables, viz. temperature, molar ratio, and catalyst loading, have independent or unrelated effects on response variable. The p-values of the interaction coefficients for "catalyst loading and temperature" and "molar ratio and temperature" are <0.05, which implies that the interaction between these variables is significant. On the other hand, p-value of interaction coefficient between catalyst and molar ratio is >0.05 that indicates insignificant interaction. The significance of regression essentially indicates that 99.80% of the effect on the triglyceride conversion is explained by the variation in the process variables or process parameters used in the statistical experimental design. Table 5.3b shows the ANOVA for the quadratic model. The Lack of Fit F-value of 3.36 and p-value of 0.105 denotes that Lack of Fit is insignificant, when compared to pure error, which also corroborates best fit of the model to data [6, 7].

A physical interpretation of the results of statistical analysis can be given as follows: With heterogeneous catalyst of KI/ZnO, the reaction system becomes 3-phase heterogeneous, viz. methanol (aqueous), oil (organic), and catalyst (solid). This system has severe mass transfer limitations that restrict interaction between the molecules. The high viscosity of oil and immiscibility of oil with methanol in mechanical stirring results in limited interfacial area that gives lower conversion i.e., 23% in 1 h of reaction. Sonication or ultrasound irradiation overcomes this limitation through micro-turbulence and micro-convections, which creates fine dispersion of the methanol and oil phases into each other. This results in generation of large interfacial area. Emulsion formation was promoted by micro-streaming, micro-convection, and acoustic shockwaves. Higher viscosity of oil than methanol gives higher convection generation in methanol phase. The micro-streaming velocity does not dependents on the temperature of the reaction medium. However, the intensity of transient bubble motion reduces with temperature of the medium, as the vapor of bulk liquid present in the bubble at the moment of transient collapse "cushions" the collapse. The maximum biodiesel yield at 335.5 K (which is near to boiling point of methanol) in this study signifies that the role of cavitation bubble in transesterification reaction is rather negligible, and intrinsic kinetics dominates the overall reaction system. Large value of activation energy supports the hypothesis of intrinsic kinetics due to the 3-phase heterogeneity present in the transesterification system, as said earlier. The molar ratio of alcohol to oil predicted by the analysis of statistical design as 10.25:1, which is much higher as compared to homogeneous catalyzed transesterification system, this also supports the mass transfer limitation in the reaction system. Lower alcohol to oil molar ratio results in higher volume fraction of oil in the reaction mixture yields low-micro-convection intensity. Low micro-convections cause not only uneven distribution of catalyst particles but also smaller interfacial area. When methanol is in large excess quantity (very high molar ratio) give rise to very small interfacial area due to very less volume of oil as compared to methanol in reaction mixture [18]. Strong interaction between all reaction parameters is predicted by the statistical model resulting in variation in biodiesel yield. The interaction parameter catalyst loading-temperature is associated through intrinsic kinetics of the reaction. The interaction parameters catalyst loading-temperature, and temperature-molar ratio are related to the alteration in convection level through the reaction medium (which determines the interfacial area) which is ultimately function of acoustic wave phenomena and cavitation bubble dynamics.

5.4 Conclusion

Potassium iodide impregnated on zinc oxide synthesised in this work has shown significant catalytic activity for transesterification reaction of soybean oil. The activity of KI/ZnO catalyst for carrying transesterification reaction is justified by the creation of K_2O phase and KI salts, confirmed through XRD analysis. They revealed a prominent catalytic activity, promoting to the high conversion of methyl esters. The superior activity is co-related to the existence of the basic sites, probably K_2O formed during the calcination. The reaction system has strong mass transfer limitations due to 3-phase heterogeneity. Large activation energy of the reaction verifies the dominance of intrinsic kinetics. The limitations have been overcome with application of ultrasound. Ultrasound increases the mass transfer by the intense mixing through micro-turbulence. The highest yield of 94.71% was obtained with just 40 min of sonication under optimized reaction conditions. The optimized reaction conditions have been determined as: catalyst concentration = 6.15 wt%; methanol to oil molar ratio = 10.25:1 and temperature = 335.5 K. All the parameters have significant effect on triglyceride conversion or biodiesel yield, and some parameters are interdepended as predicted by the statistical analysis of experimental design. The catalyst showed good retention of activity (up to 50%) after five successive cycles. We believe that findings of our study will be beneficial for future research in this area, and the methodology and catalyst developed in present study can also be stretched to non-edible oils reaction system.

Acknowledgements Authors acknowledge Analytical and Petroleum Laboratory, Chemical Engineering Department of IIT Guwahati and also Analytical Facilities provided by Central Instrumental Facility of IIT Guwahati.

References

1. Alonso DM, Mariscal R, Granados ML, Maireles-Torres P (2009) Biodiesel preparation using Li/CaO catalysts: activation process and homogeneous contribution. Catal Today 143:167–171
2. Aransiola EF, Ojumu TV, Oyekola OO, Madzimbamuto TF, Ikhu-Omoregbe DIO (2014) A review of current technology for biodiesel production: state of the art. Biomass Bioenergy 61:276–297
3. Choudhury HA, Chakma S, Moholkar VS (2014) Mechanistic insight into sonochemical biodiesel synthesis using heterogeneous base catalyst. Ultrason Sonochem 21:169–181
4. Choudhury HA, Goswami PP, Malani RS, Moholkar VS (2014) Ultrasonic biodiesel synthesis from crude Jatropha oil with heterogeneous base catalyst: mechanistic insight and statistical optimization. Ultrason Sonochem 21:1050–1064
5. Choudhury HA, Malani RS, Moholkar VS (2013) Acid catalyzed biodiesel synthesis from Jatropha oil: mechanistic aspects of ultrasonic intensification. Chem Eng J 231:262–272
6. Dharma S, Masjuki HH, Ong HC, Sebayang AH, Silitonga AS, Kusumo F, Mahlia TMI (2016) Optimization of biodiesel production process for mixed *Jatropha curcas–Ceiba pentandra* biodiesel using response surface methodology. Energy Convers Manage 115:178–190
7. El-Gendy NSh, Deriase SF, Hamdy A, Abdallah RI (2015) Statistical optimization of biodiesel production from sunflower waste cooking oil using basic heterogeneous biocatalyst prepared from eggshells. Egypt J Pet 24:37–48
8. Evangelista JPC, Chellappa T, Coriolano ACF, Fernandes VJ Jr, Souza LD, Araujo AS (2012) Synthesis of alumina impregnated with potassium iodide catalyst for biodiesel production from rice bran oil. Fuel Process Technol 104:90–95
9. Kalva A, Sivasankar T, Moholkar VS (2009) Physical mechanism of ultrasound-assisted synthesis of biodiesel. Ind Eng Chem Res 48:534–544
10. Kumar M, Sharma MP (2015) Assessment of potential of oils for biodiesel production. Renew Sustain Energy Rev 44:814–823
11. Liu H, Su LY, Shao Y, Zou LB (2012) Biodiesel production catalyzed by cinder supported CaO/KF particle catalyst. Fuel 97:651–657
12. Liu XJ, He HY, Wang YJ, Zhu SL, Piao XL (2008) Transesterification of soybean oil to biodiesel using CaO as a solid base catalyst. Fuel 87:216–221
13. Lukic I, Krstic J, Jovanovic D, Skala D (2009) Alumina/silica supported K2CO3 as a catalyst for biodiesel synthesis from sunflower oil. Biores Technol 100:4690–4696
14. Luu PD, Takenaka N, Luu BV, Pham LN, Imamura K, Maeda Y (2014) Co -solvent method produce biodiesel form waste cooking oil with small pilot plant. Energy Procedia 61:2822–2832
15. Moholkar VS, Choudhary HA, Singh S, Khanna S, Ranjan A, Chakma S, Bhasarkar JB (2015) Physical and chemical mechanisms of ultrasound in biofuel synthesis. In: Fang Z, Smith RL, Jr, Qi X (eds) Biofuels and biorefineries, vol.4. Production of biofuels and chemicals with ultrasound. Spinger, Berlin, pp 35–86
16. Mootabadi H, Salamatinia B, Bhatia S, Abdullah AZ (2010) Ultrasonic-assisted biodiesel production process from palm oil using alkaline earth metal oxides as the heterogeneous catalysts. Fuel 89:1818–1825
17. Murray PM, Bellany F, Benhamou L, Bučar D-K, Taborb AB, Sheppard TD (2016) The application of design of experiments (DoE) reaction optimisation and solvent selection in the development of new synthetic chemistry. Org Biomol Chem 14:2373–2384
18. Shah YT, Pandit AB, Moholkar VS (1999) Cavitation reaction engineering. Plenum Press, New York
19. Takase M, Zhang M, Feng W, Chen Y, Zhao T, Cobbina SJ, Yang L, Wu X (2014) Application of zirconia modified with KOH as heterogeneous solid base catalyst to new non-edible oil for biodiesel. Energy Convers Manage 80:117–125

20. Tang Y, Meng M, Zhang J, Lu Y (2011) Efficient preparation of biodiesel from rapeseed oil over modified CaO. Appl Energ 88:2735–2739
21. Wen LB, Wang Y, Lu DL, Hu SY, Han HY (2010) Preparation of KF/CaO nanocatalyst and its application in biodiesel production from Chinese tallow seed oil. Fuel 89:2267–2271
22. Xie W, Li H (2006) Alumina-supported potassium iodide as a heterogeneous catalyst for biodiesel production from soybean oil. J Mol Catal A: Chem 255:1–9
23. Yang Z, Xie W (2007) Soybean oil transesterification over zinc oxide modified with alkali earth metals. Fuel Proc Technol 88:631–638
24. Zabeti M, Daud W, Aroua MK (2010) Biodiesel production using alumina-supported calcium oxide: an optimization study. Fuel Proc Technol 91:243–248

Chapter 6
Experimental Evaluation of Cottonseed Biodiesel as an Alternative Fuel for Diesel Engine

Pankaj S. Shelke, Nitin M. Sakhare, Subhash Lahane and N.G. Patil

Abstract Diesel engines are commonly used due to its low fuel consumption and higher torque. Nowadays, the world is facing the challenges of crises of fossil fuels due to rapid increase in utilization of its in transportation and industrialization. This further leads to environmental degradation by polluting the air. Hence, in the modern world, more importance is given on alternative source of energy. Several possible alternative sources are available as, namely biodiesel, alcohols, CNG, LPG, producer gas, etc. Biodiesel presents a promising alternative to diesel fuel since it is renewable and produced easily by transesterification process. Therefore, the research work is aimed at experimental evaluation of cottonseed biodiesel as an alternative fuel for diesel engine. Cottonseed is non-edible oil, thus food versus fuel conflict will not arise if this is used for biodiesel production. Biodiesel blends (B5, B10, B15, and B20) are prepared using cottonseed biodiesel and diesel at laboratory scale. The performance and emission characteristics of biodiesel blends fueled diesel engine are compared with base diesel. It is found that CO, HC, and smoke emissions are decreased with all biodiesel blends than base diesel. The cleaner and complete combustion take place due to oxygen content in biodiesel which helps to reduce CO and HC emissions. However, brake-specific fuel consumption (BSFC) and NO_x emission are increased with biodiesel blends than base diesel. BSFC increased due to lower calorific value of biodiesel, whereas NO_x emission increased due to advance in dynamic injection timing due to higher bulk modulus, higher spray penetration, higher in-cylinder temperature and oxygen content in biodiesel. The brake power is decreased from 3 kW with base diesel to 2.93, 2.9, 2.79, and 2.77 kW with B5, B10, B15, and B20, respectively.

Keywords Cottonseed biodiesel · Performance · Emission · Combustion
Diesel engine

P.S. Shelke · N.M. Sakhare · S. Lahane (✉) · N.G. Patil
Department of Mechanical Engineering, MIT Aurangabad, Aurangabad
Maharashtra, India
e-mail: subhashlahane@gmail.com

S. Kumar et al. (eds.), *Conference Proceedings of the Second International Conference on Recent Advances in Bioenergy Research*, Springer Proceedings in Energy, https://doi.org/10.1007/978-981-10-6107-3_6

6.1 Introduction

The manufacturer and developer of engine play a leading role in the field of power, propulsion, and energy. The purpose of IC Engine is the production of mechanical power from chemical energy of the fuel [1]. The major problems of using fossil fuels are increase in greenhouse gases, global warming, weakening of ozone layer, and acid rain [2]. Due to these disadvantages of fossil fuels, the researches take interest in developing alternative fuels for IC Engine. There are various alternative sources of fuels like vegetable oil, biogas, biomass, and alcohol which are all renewable in nature. Among these, vegetable oil is a very important alternative to diesel fuel for internal combustion engines because it is widely available, biodegradable, non-toxic, and environmental friendly. In agriculture-based country like India, the use of vegetable oil has to be identified and indicated in order to prevent environmental degradation [3]. The utilization of vegetable oils in internal combustion engines is not a recent innovation. The Rudolf Diesel used peanut vegetable oil in his diesel engine [4]. Transesterification is the process of separating the fatty acid from the glycerol backbone to form free glycerol and fatty acid esters (FAE) which is commonly known as biodiesel [5]. Biodiesel contains 10–11% oxygen by weight, has high cetane number than diesel, and has environmental benefits such as lower emission of CO, CO_2, and unburned hydrocarbon (UBHC) [6, 7]. The fuel properties of cottonseed oil seem to meet the fundamental requirement of diesel engine. This is the main reason that no much modification is required to use biodiesel in diesel engine [8]. Capered et al. [9] revealed that by using B5, B20, and B100, rated power of 13.6, 13.4, and 13.1 kW is produced. Brake-specific fuel consumption increases by using B5, B20, and B100. Carbon monoxide emissions decreased by an average 15 and 19% by using B5 and B100, respectively. Hydrocarbon emissions decreased by 14 and 26% by using B5 and B100, respectively. NO_x emission is decreased by 4, 5, and 14% with B5, B20, and B100, respectively. Xue et al. [10] revealed that the use of biodiesel in diesel engine reduced PM, HC, and CO emissions, whereas NO_x emissions increased with the imperceptible power loss. In case of biodiesel, the fuel consumption increases than diesel. The NO_x emission increases due to higher oxygen content for biodiesel than diesel. The use of biodiesel blends in place of diesel helps to control air pollution without much-sacrificing engine power and economy. Huang et al. [11] burning efficiency improved due to high oxygen content in biodiesel. Emissions like CO, HC, and PM decreases but NO_x emissions increases due to high temperature burning environment [11]. Powell et al. [12] reported that biodiesel is a clean burning alternative renewable fuel produced from plant oils, and it is easily blended with diesel to make biodiesel blends. Sani et al. [13] observed that the brake power decreased by 4–5% and brake-specific fuel consumption increased by 5–10% with cottonseed biodiesel than diesel. NO_x emission is increased by 11–22% than diesel. Karabektas et al. [7] used preheated cottonseed biodiesel in diesel engine. The brake power decreases by 1.92 and 7.59% with B10 and B20 than diesel fuel. Brake thermal efficiency is higher with B20 and B10 as compared to diesel fuel. CO emission is decreased by 14.40–45.66% with COME (cottonseed oil methyl ester) than diesel fuel. The NO_x emissions increased by 11.21–39.1% with COME as compared to diesel fuel. Aydin et al. [14] observed slight reductions in the engine torque and power due to lower heating value of CSOME (cottonseed oil methyl ester) than diesel fuel. The CO emissions decreased with biodiesel than diesel due to higher oxygen content. In dissimilarity to many researchers, the NO_x emission was decreased for all blends and B100 except for B5 in their experiments [15].

6.2 Experimental Details

The setup consists of single cylinder, four strokes, and diesel engine connected to eddy current dynamometer as shown in Fig. 6.1. The setup consists of various measuring instrument to measure torque, air flow rate, fuel consumption, combustion pressure, crank angle, and load measurement. These signals are interfaced to computer through high-speed data acquisition device.

The engine specification on which the experimentation is carried out for different biodiesel blends at various load is given in Table 6.1. The properties of the diesel and cottonseed biodiesel are tabulated in Table 6.2; these properties are compared with ASTM standards. The properties were tested at Indian biodiesel corporation Baramati, Maharashtra, India.

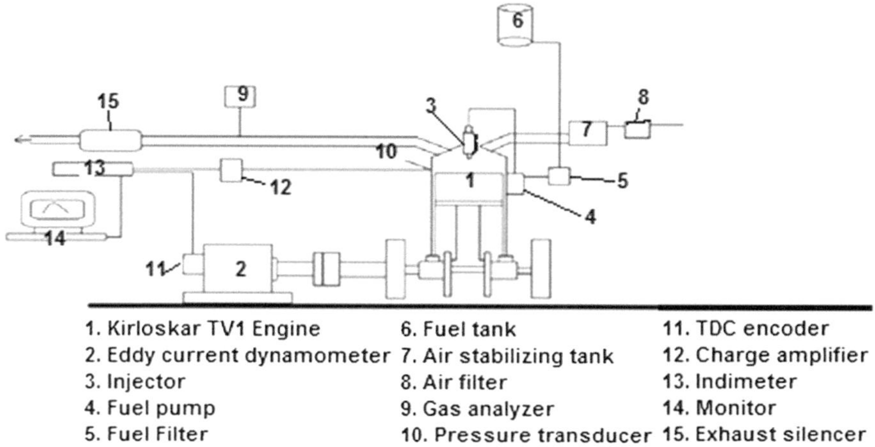

Fig. 6.1 Experimental setup

Table 6.1 Engine specifications

No. of cylinder	1
No of strokes	4
Fuel	H.S. Diesel
Rated power	3.5 kW @ 1500 rpm
Cylinder diameter	87.5 mm
Stroke length	110 mm
Connecting rod length	234 mm
Compression ratio	12:1 to 18:1
Nozzle diameter	0.2 mm
Dynamometer arm length	185 mm

Table 6.2 Properties of cottonseed biodiesel and diesel

Property	Diesel	Cottonseed biodiesel blends			
		B5	B10	B15	B20
Density (kg/m^3)	825	836	838	842	844
Calorific value (MJ/kg)	42.5	42.36	42.16	41.9	41.55
Cetane number	49.5	–	–	49.87	–
Viscosity (cSt)	2.7	–	–	3.2	–
Flash point (°C)	64	–	–	122	–
Fire point (°C)	71	–	–	122	–

6.3 Results and Discussion

The comparison of brake power (BP) for different biodiesel blends with diesel fuel is represented in Fig. 6.2a. The brake power was calculated using Eq. 6.1. The experiments were conducted on a single-cylinder diesel engine using cottonseed biodiesel and diesel as a fuel. In this experiment, load is varied from 25 to 100% with a step of 25% for base diesel (B0) and biodiesel blends (B5, B10, B15, and B20). The graph between brake power and load is compared in Fig. 6.2a. The brake power for different cottonseed biodiesel blends is just lower than the diesel fuel. The brake power decreases from 3 kW with diesel to 2.93, 2.9, 2.79, and 2.77 kW at the rated load (100%) with B5, B10, B15, and B20 biodiesel blends, respectively. It is due to the lower calorific value (CV) of biodiesel blends than diesel.

Figure 6.2b illustrates the comparison of brake thermal efficiency (BTE) with the change in load. The brake thermal efficiency is calculated using Eq. 6.2. The BTE of cottonseed biodiesel decreases than that of the diesel fuel. The BTE decreases from 28.86% with diesel to 28.59, 27.85, 27.19, and 27.05% at the rated load (100%) with B5, B10, B15, and B20 biodiesel blends, respectively. The factors like lower heating values and higher viscosity of the biodiesel may affect the mixture formation process which have the adverse effect on brake thermal efficiency for biodiesel blends as compared to base diesel. It is also shown in Fig. 6.3a that the

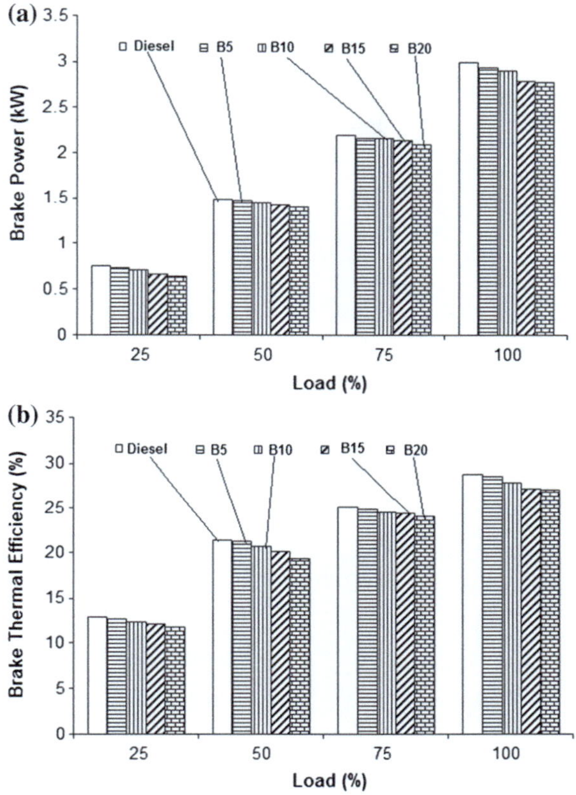

Fig. 6.2 Comparison of **a** brake power and **b** brake thermal efficiency for diesel and biodiesel blends

cumulative heat release rate is decreased with cottonseed biodiesel blends. This may be one of the reasons for the decrease in brake thermal efficiency. The specific heat at constant pressure and volume, and gama is used to calculate heat release rate is shown in Fig. 6.3b. Another reason for the decrease in brake thermal efficiency may be due to lower mass burning fraction rate. The mass burning fraction rate is lower in case of biodiesel blends as compared to base diesel as shown in Fig. 6.4a. The mass burned fraction (x_b) is calculated by using Eq. 6.3. The density ratio (ρ_u/ρ_b) does not depend on the equivalence ratio, burned gas fraction in the unburned mixture, gas temperature, and pressure, its value is close to 5. The volume fraction (y_b) can be calculated by using Eq. 6.4. The biodiesel blend has higher density, viscosity, and surface tension which influence the fuel spray characteristics such as spray cone angle, Sauter mean diameter, spray penetration, and vaporization of biodiesel blends. Sauter mean diameter is higher with biodiesel blends which take more time to vaporize. As the latent heat of vaporization is higher with biodiesel blends, it deteriorates the vaporization rate.

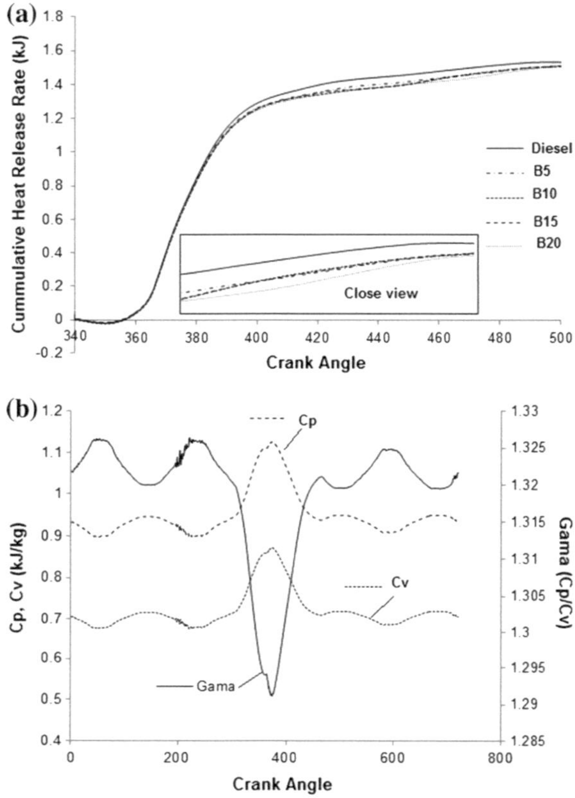

Fig. 6.3 Comparison of **a** cumulative heat release rate and **b** Cp, Cv, and gama for diesel and biodiesel

$$\text{BP} = \frac{2\pi \times N \times T}{60{,}000} \tag{6.1}$$

$$\text{BTE} = \frac{BP}{\dot{m_f} \times CV} \tag{6.2}$$

$$x_b = \left[1 + \frac{\rho_u}{\rho_b}\left(\frac{1}{y_b} - 1\right)\right]^{-1} \tag{6.3}$$

$$y_b = \frac{V_b}{V} \tag{6.4}$$

Figure 6.4b represents the specific fuel consumption (SFC) with respect to load for diesel and biodiesel blends. It is concluded that as the load increases the fuel consumption also increase. The fuel consumption in case of cottonseed biodiesel blends is higher than that of diesel. It is due to lower calorific value of biodiesel

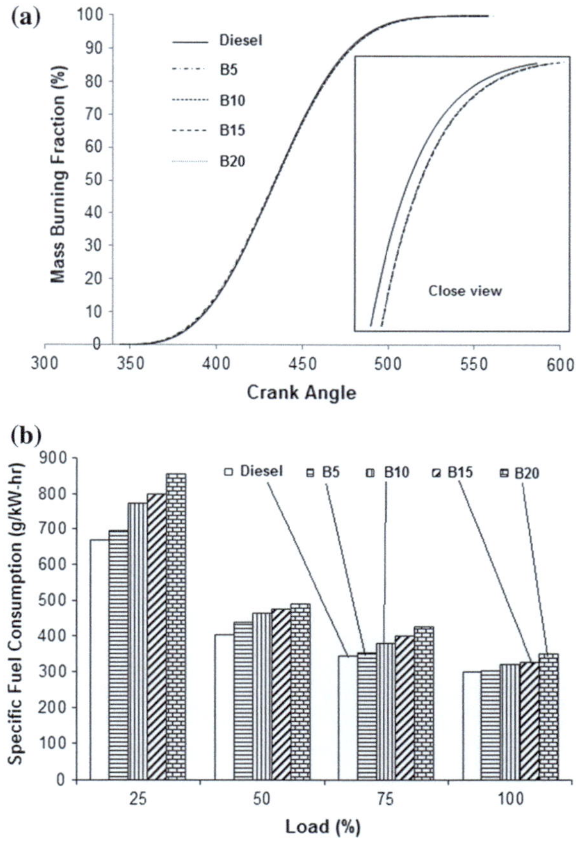

Fig. 6.4 Variation of **a** mass burning fraction and **b** SFC for diesel and biodiesel blends

than diesel. The fuel consumption increases from 298.3 g/kW-h with diesel to 301.8, 319.3, 326.3, and 347.3 g/kW-h with B5, B10, B15, and B20 biodiesel blends, respectively, at the rated load. This may be another reason for the decrease in brake thermal efficiency in case of biodiesel blends.

Figure 6.5a illustrates the variation in brake mean effective pressure (BMEP) with the change in load. Mean effective pressure BMEP is defined as the hypothetical pressure which is thought to be acting on the piston throughout the power stroke. Cottonseed biodiesel blends give lower brake mean effective pressure than diesel hence engine operates smoothly. The brake mean effective pressure decreases slightly from 3.47 bar with diesel to 3.46, 3.46, 3.41, and 3.39 bar with B5, B10, B15, and B20 biodiesel blend, respectively, at the rated load. The rate of pressure rise (RPR) also decreased with all biodiesel blends as compared to base diesel as shown in Fig. 6.5b, which confirms the smooth running operation of biodiesel

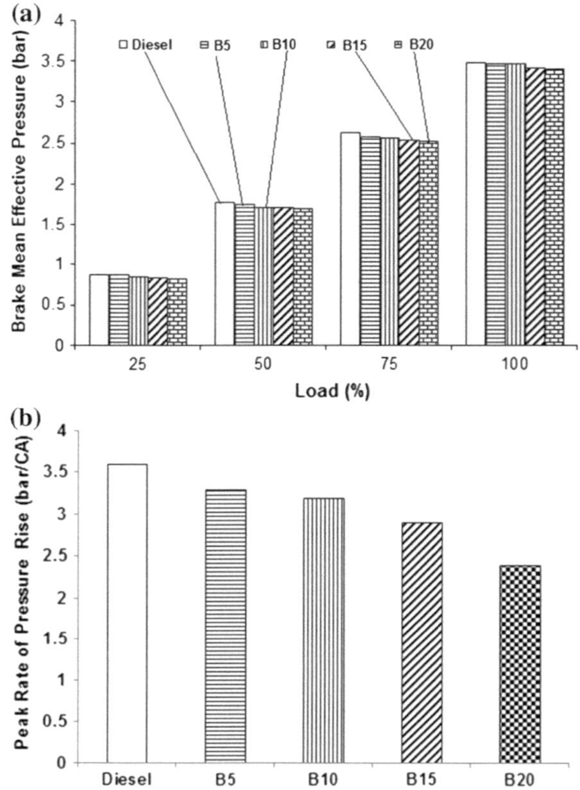

Fig. 6.5 Comparison of **a** brake mean effective pressure and **b** peak RPR for diesel and biodiesel blends

fueled diesel engine. The rate pressure rise decreased mainly due to higher cetane number of biodiesel results in shorter ignition delay.

CO emissions in case of cottonseed biodiesel decreased due to higher oxygen content (Fig. 6.6a). Due to higher oxygen content, most of the CO emissions are oxidized to form CO_2. It is found that the percentage change in CO emission is −10.6, −27.3, −44, and −60 for B5, B10, B15, and B20 biodiesel blends as compared to base diesel at the rated load. HC emission in case of cottonseed biodiesel decreased due to higher oxygen content results in cleaner combustion, higher in-cylinder pressure and temperature (Fig. 6.6b). Due to higher oxygen content, complete combustion takes place with high in-cylinder temperature hence HC emissions decreased. The higher in-cylinder temperature and oxygen content is responsible for higher NO_x emission which is discussed in later. It is found that HC emission is decreased from 9 ppm with diesel to 7, 6, 4, and 2 ppm for B5, B10, B15, and B20 biodiesel blends at the rated load.

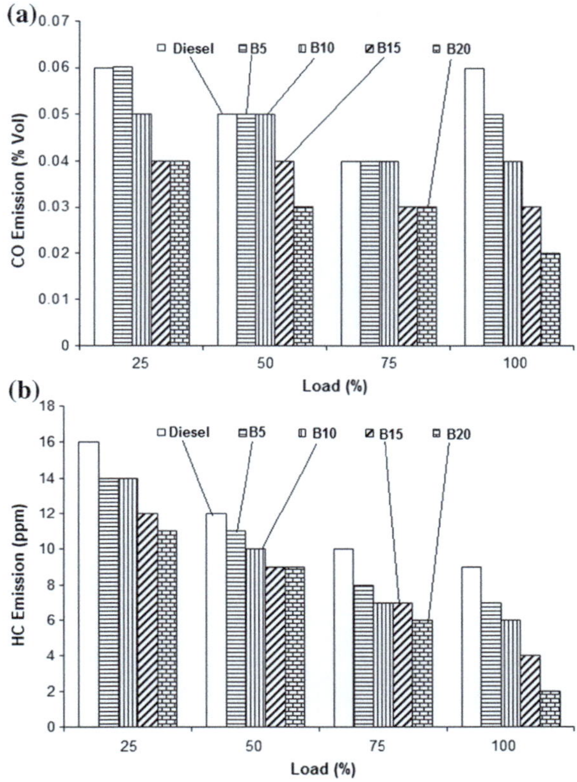

Fig. 6.6 Comparison of **a** CO emission and **b** HC emission for diesel and biodiesel blends

CO_2 emission increased with cottonseed biodiesel blends as compared to diesel fuel as shown in Fig. 6.7a. The CO_2 emission is helpful in the plants photosynthesis process. Hence, the biodiesel would not emit extra amount of CO_2 to the atmosphere. It is found that CO_2 emissions of B5, B10, B15, and B20 biodiesel are increased 1.47, 1.75, 1.77, and 1.91% vol. from 1.01% vol. with diesel at rated load condition. NO_x emission increases with the biodiesel than the diesel fuel (Fig. 6.7b). High oxygen content in biodiesel favors a high in-cylinder temperature and high premixed combustion phase results in higher NO_x emission. NO_x emission increased from 215 ppm with diesel to 218, 219, 222, and 225 ppm with B5, B10, B15, and B20, respectively, at the rated load condition. The another reason for higher NO_x emission is an automatic advance in dynamic injection timing [15], higher in-cylinder temperature and oxygen content [16, 17], and higher fuel spray penetration [18].

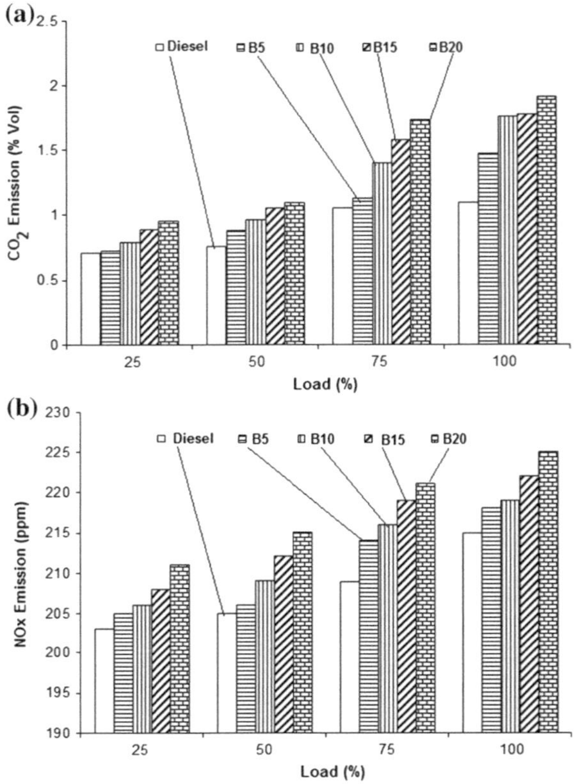

Fig. 6.7 Comparison of **a** CO_2 emission and **b** NO_x emission for diesel and biodiesel blends

6.4 Conclusions

The following conclusions are drawn based on the experimental results of the diesel engine with cottonseed biodiesel blends as compared to base diesel.

- The brake thermal efficiency decreased with biodiesel from 28.86% with the base diesel to 28.59, 27.85, 27.19, and 27.05% with B5, B10, B15, and B20 biodiesel blends, respectively, at the rated load. It is due to lower calorific value fuel and lower cumulative heat release rate and higher specific fuel consumption.
- CO and HC emissions are decreased with biodiesel content as compared to base diesel. HC emission is decreased from 9 ppm with diesel to 7, 6, 4, and 2 ppm for B5, B10, B15, and B20 biodiesel blends at the rated load.
- CO_2 and NO_x emissions are increased with biodiesel content as compared to base diesel. NO_x emission increased from 215 ppm with diesel to 218, 219, 222, and 225 ppm with B5, B10, B15, and B20, respectively, at the rated load

condition. It is due to higher bulk modulus of biodiesel results in automatic advance in injection timing, higher fuel spray penetration, higher in-cylinder temperature and oxygen content in the fuel.

References

1. Haywood JB (1988) Internal combustion engine. McGraw-hill, New York
2. Shahid E, Jamal Y (2011) Performance evaluation of a diesel engine using biodiesel. Pak J Eng Appl Sci 9:68–75
3. Kuthalingam A, Asokan G (2013) Performance and emission characteristic of double biodiesel blend with diesel. Therm Sci 17(1):255–262
4. Augustine A, Marimuthu L, Muthusam S (2012) Performance and evaluation of DI diesel by using preheated cottonseed oil methyl ester. International conference on modeling optimization and computing, Procedia Engineering vol 38, pp 779–790
5. Upadhyay Y, Sharma R, Pal A (2013) Production of biodiesel from non edible cottonseed oil by mechanical stirrer technique. IOSR J Mech Civil Eng 8(1):22–26
6. Fan X, Wang X, Chen F, Geller D, Wan P (2008) Engine performance test of cottonseed oil biodiesel. Open Energy Fuels J 40–45
7. Karabektas M, Ergen G, Hosoz M (2008) The effects of preheated cottonseed oil methyl ester on the performance and exhaust emissions of a diesel engine. Appl Therm Eng 28:2136–2143
8. He Y, Bao Y (2005) Study on cottonseed oil as a partial substitute for diesel oil in fuel for single-cylinder diesel engine. Renew Energy 30:805–813
9. Capered S, Powell J, Parnell C (2008) Engine performance and exhaust emissions of cottonseed oil biodiesel. Beltwide Cotton Conferences, Nashville, Tennessee
10. Xue J, Grift TE, Hansen AC (2011) Effect of biodiesel on engine performances and emissions. Renew Sustain Energy Rev 15:1098–1116
11. Huang D, Zhou H, Lin L (2012) Biodiesel: an alternative to conventional fuel. International conference on future energy, environment, and materials, Energy Procedia vol 16, pp 1874–1885
12. Joseph J, Sergio P, Parnell (2007) Diesel engine performance and exhaust emissions using cottonseed oil biodiesel. Beltwide cotton conferences, New Orleans, Louisiana
13. Sani F, Abdulmalik I, Rufai I (2013), Performance and emission characteristics of compression ignition engines using biodiesel as a fuel: a review. Asian J Nat Appl Sci
14. Aydin H, Bayindir H (2010) Performance and emission analysis of cottonseed oil methyl ester in a diesel engine. Renew Energy 35:588–592
15. Subramanian KA, Lahane S (2013) Comparative evaluations of injection and spray characteristics of a diesel engine using karanja biodiesel–diesel blends. Int J Energy Res 37 (6):582–597
16. Lahane S, Subramanian KA (2015) Effect of different percentages of biodiesel–diesel blends on injection, spray, combustion, performance, and emission characteristics of a diesel engine. Fuel 139:537–545
17. Subramanian KA, Lahane S (2012) Comparative assessment of injection, combustion, performance and emission characteristics of a diesel engine for biodiesel–diesel blends. Int J Renew Energy Technol 3(4):410–429
18. Lahane S, Subramanian KA (2014) Impact of nozzle holes configuration on fuel spray, wall impingement and NOx emission of a diesel engine for biodiesel–diesel blend (B20). Appl Therm Eng 64(1–2):307–314

Part III
Thermochemical

Chapter 7
Pyrolytic Characterization and Kinetic Analysis of *Camellia sinensis* (Tea) Seed Deoiled Cake

Nabajit Dev Choudhury, Bichitra Bikash and Rupam Kataki

Abstract The knowledge of kinetic parameters is essential for modeling of the reactor and for optimization of the process conditions. In the present work, the first attempt was made to investigate the thermal behavior and to calculate the kinetic parameters of *Camellia sinensis* deoiled cake (CSDC) using thermogravimetric analysis. The pyrolysis experiments were conducted in temperatures ranging from 298 to 1273 K under inert atmosphere utilizing two different heating rates of 10 and 40 K min^{-1}, respectively. On the basis of thermogravimetric analysis, activation energy (*E*), pre-exponential factor (*A*) for the active pyrolysis zone of the deoiled cake were calculated, respectively, by Arrhenius, Coast-Redfern and Flynn-Wall-Ozawa methods for the heating rates of 10 and 40 K min^{-1}. These results can provide useful information to predict kinetic model of CSDC pyrolysis and optimization of the process conditions. The activation energy was found in the range of 54–77 kJ mol^{-1} for heating rates of 10 and 40 K min^{-1}, respectively. The results derived are useful for preliminary assessment of *C. sinensis* seed deoiled cake as potential feedstock for thermal conversion.

Keywords Pyrolysis · Deoiled cake · TGA · Kinetics

7.1 Introduction

The epic challenge of twenty-first century is to meet the mankind's up surging energy demand with the clean, reliable, and inexpensive energy sources. Moreover, the depleting nature of conventional fossil fuels coupled with economical and environmental concern issues has compelled the scientific endeavors for exploring and exploiting new and renewable biobased feedstocks toward energy generation.

N.D. Choudhury (✉) · B. Bikash
Assam Down Town University, Guwahati 781026, Assam, India
e-mail: nabajit2013@gmail.com

R. Kataki
Department of Energy, Tezpur University, Napaam, Tezpur 784028, Assam, India

S. Kumar et al. (eds.), *Conference Proceedings of the Second International Conference on Recent Advances in Bioenergy Research*, Springer Proceedings in Energy, https://doi.org/10.1007/978-981-10-6107-3_7

Harnessing energy from biomass resources is one of the prospects with massive potential to meet increasing energy demand. Harnessing energy from biomass resources is not new in practice, and numerous scientific literature pertaining to bioenergy feedstocks are available.

The biofuels (biogas, biodiesel, and bioethanol) derived from biomass are considered as a good alternative to petroleum fuels. In an agriculture-based economy like India, the prospect of utilizing non-edible oil seeds for biofuel generation is a viable alternative to conventional fossil fuels. This view is further supported by the Government of India's recent biofuel policy, which targets bringing more and more waste, unused lands, and open forest areas for plantation of tree/shrub species yielding non-edible oil seeds to produce commercial grade biofuel. Considering the future scenario of non-edible oil seed utilization, there is a need for efficient utilization of their deoiled cakes. In recent time due to increasing demand of biodiesel, lots of oil cakes have increased tremendously and about 2 tonnes of oil cake are dumped as a waste for every tonne of biodiesel production [1]. In this regard, the present investigation targets *Camellia sinensis* deoiled cake (CSDC, a biowaste, produced in the process of biodiesel production) as a potential candidate for bioenergy generation. This work reports the systematic investigation of the physicochemical properties of the CSDC, its decomposition behavior (activation energy, exponential factor), and kinetics using Arrhenius, Coats–Redfern and Flynn-Wall-Ozawa methods at different heating rates (10 and 40 K min^{-1}).

7.2 Materials and Methods

In this study, the samples (Fig. 7.1) were collected from Biomass Conversion and the Gasification Laboratory of Department of Energy, Tezpur University, India, after lipid extraction with mechanical oil expeller [Malnad type oil expeller (Indus)]. The samples were grounded using a Wiley mill and sieved to a particle size of 0.4 mm (40 mesh) screen (as per TAPPI T257 Om-85 methods) and then samples were oven dried till constant weight and kept in a desiccator. The moisture content, ash content, and volatile matter of *C. sinensis* deoiled cake (CSDC) were determined according to ASTM D 3173-75. The ultimate analysis was done using the CHN analyzer (Perkin Elmer, 2400 Series-II). The percentage of oxygen was determined by means of difference: Oxygen = 100 − (Carbon + Hydrogen + Nitrogen). Gross calorific value (GCV) was determined using an automatic adiabatic bomb calorimeter (Changsha Kaiyuan Instruments Co, 5E-1AC/ML). The thermogravimetric (TG) analysis was carried out in Pyris diamond TG/differential thermogravimetric analyzer (Perkin Elmer). Experiments were conducted non-isothermally at two different heating rates of 10 and 40 K min^{-1} in N_2 environment.

Fig. 7.1 *Camellia sinensis* and CSDC

7.3 Thermal Analysis and Calculation of Kinetic Parameters

7.3.1 *Kinetic Modeling*

Thermogravimetric analysis is the most commonly used technique to study the reaction kinetics of the pyrolysis process. The pyrolysis of biomass is frequently represented by the following reaction scheme:

$$\text{Biomass solid} \rightarrow \text{volatiles} + \text{Solid residues}$$

The fraction of pyrolyzed biomass (conversion factor), α, is defined by the following expression:

$$\alpha = \frac{(m_0 - m_t)}{(m_0 - m_\infty)} \tag{7.1}$$

where m_0, m_t, and m_∞ refer to the values of biomass at the beginning, at time t and at the end of the degradation event of interest, respectively.

Pyrolysis rate, $\mathrm{d}\alpha/\mathrm{d}t$, is a linear function of temperature dependent rate constant, k, and $f(\alpha)$, a temperature-independent function of conversion:

$$\frac{\mathrm{d}\alpha}{\mathrm{d}t} = kf(\alpha) \tag{7.2}$$

Replacing the rate constant with Arrhenius equation, and introducing the non-isothermal condition, heating rate ($\beta = \mathrm{d}T/\mathrm{d}t$), Eq. 7.2 becomes:

$$\frac{\mathrm{d}\alpha}{\mathrm{d}t} = \frac{A}{\beta}\mathrm{e}^{\left(\frac{E}{RT}\right)}f(\alpha) \tag{7.3}$$

where A is the pre-exponential factor, E is the activation energy, R is the gas constant, and T is the absolute temperature.

Selecting nth order reaction model in the light of a previously accomplished kinetics study in the literature [2] and rearranging, Eq. 7.3 becomes:

$$\frac{\mathrm{d}\alpha}{(1-\alpha)^n} = \frac{A}{\beta} e^{\left(\frac{E}{RT}\right)} f(\alpha) \tag{7.4}$$

In this study, Eq. 7.4 is the fundamental expression used in model-fitting kinetic calculation methods on the basis of TG data.

7.3.2 Arrhenius Method

The final rate equation of Arrhenius method can be obtained by taking the logarithm of Eq. 7.4 and making some rearrangements. It is given as follows:

$$\ln\frac{\mathrm{d}\alpha}{\mathrm{d}T} - n\ln(1-\alpha) = \ln\left(\frac{A}{\beta}\right) - \frac{E}{RT} \tag{7.5}$$

According to Eq. 7.5, the plot of $\ln\frac{\mathrm{d}\alpha}{\mathrm{d}T} - n\ln(1-\alpha)$ versus $\frac{1}{T}$ should give a straight line for the appropriate value of reaction order n. The activation energy and pre-exponential factor of each active pyrolysis stages were calculated from the related slope $(-E/R)$ and interception $\ln(A/\beta)$ of final plots, respectively.

7.3.3 Coats–Redfern Method

In Coats–Redfern method, the integral of Eq. 7.4 is taken, and the resulting exponential integral which does not have an exact analytical solution is approximated using a Taylor series expansion. The obtained equation is simplified by considering $2RT/E \ll 1$ [3]. The final form is given as follows:

$$\ln g(\alpha) = -\frac{E}{RT} + \ln\left(\frac{AR}{\beta E}\right) \tag{7.6}$$

where $g(\alpha) = \frac{-\ln(1-\alpha)}{T^2}$ if $n = 1$; $g(\alpha) = \frac{(1-\alpha)^{1-n}}{(1-n)T^2}$ if $n \neq 1$. In this method, a straight line should be obtained by plotting $\ln g(\alpha)$ versus $(1/T)$ if n is selected properly. The activation energy and pre-exponential factor of each active pyrolysis stage were determined from the slope $\left(-\frac{E}{R}\right)$ and the interception $\ln\left(\frac{AR}{\beta E}\right)$ of final plots, respectively.

7.3.4 FWO Method

The Flynn-Wall-Ozawa (FWO) method is an integral isoconversional technique in which activation energy is related to heating rate and temperature at a constant conversion. The equation is as follows [4]:

$$\ln\beta = C_1 - \frac{E}{RT} \tag{7.7}$$

where C_1 is constant. Activation energy can be calculated from the slope of $\ln\beta$ versus $1/T$. Hence, the temperatures corresponding to the fixed values of conversion were measured at two different heating rates, and necessary plots were drawn (Fig. 7.3). Activation energy of active pyrolysis stages was calculated from the slopes $(-E/T)$ of related plots at 0.1–0.9 conversion interval.

7.4 Results and Discussion

Table 7.1 provides the biomass properties of CSDC with their mean characteristic composition. It was reported that to ensure rapid heat transfer rates in a fast pyrolysis reactor, the moisture content should be less than 10 wt% [5]. In this

Table 7.1 Properties of CSDC

Properties		*Camellia sinensis* deoiled cake	
Proximate analysis (wt%)			
Moisture		4.34 ± 0.27	
Volatile matter		80.22 ± 0.48	
Fixed Carbon		10.66 ± 0.41	
Ash		4.78 ± 0.66	
Ultimate analysis (wt%)			
Carbon		47.63	
Hydrogen		7.34	
Nitrogen		3.48	
Oxygen		41.55	
H/C molar ratio (on ash free basis)		1.849	
O/C molar ratio (on ash free basis)		0.654	
Empirical formula		$CH_{1.849}N_{0.062}O_{0.654}$	
Gross calorific value (MJ/kg)		19.65	
Biomass composition			
Cellulose	Hemicellulose	Lignin	Extractives
35.55	28.42	25.67	10.36

regard, CSDC with low moisture content of 4.61% and high volatile matters of 80.7% is a potential candidate for thermal conversion. The ash content and fixed carbon content of CSDC were 4.16 and 15.08%, respectively.

Elemental composition of C, H, O, and N in CSDC was 47.63, 7.34, 41.55, and 3.48%, respectively (Table 7.1). The empirical formula of the deoiled cake is $CH_{1.849}N_{0.062}O_{0.654}$. The H/C and O/C molar ratios (on an ash-free dry basis) were calculated from elemental composition as 1.849 and 0.654, respectively. The GCV of CSDC is found as 19.65 MJ/Kg. The higher proportion of hydrogen and oxygen compared with carbon reduces the energy value of the CSDC due to lower energy contained in carbon–oxygen and carbon–hydrogen bonds than in carbon–carbon bonds [6].

The TG and DTG curves for heating rates of 10 and 40 °C/min are shown in Fig. 7.2a, b. As shown in the figures, the entire pyrolysis process can be divided into three stages, since every single slope on a TG curve indicates the beginning of a new stage. Considering the TG/DTG curve obtained at 10 °C/min, the first stage started from 22 °C and finished at initial temperature T_i of devolatization 118 °C. The first tiny peak observed in this stage is due to the exclusion of physically adsorbed moisture in the biomass. The second stage is associated with simultaneous degradation of cellulose and hemicellulose started at 118 °C(T_i) and continues to 400 °C which is the end of the main devolatization (T_e). From the DTG curve, it is seen that maximum mass loss rate occurred at 316 °C. The temperature of maximum reaction rate is referred as T_{max}. Due to high decomposition rate, the second stage of decomposition can be treated as active pyrolysis zone. The high reaction rate at this stage led to the formation of pyrolysis products. The third stage starts at 400 °C and continues up to 1000 °C, but the rate of degradation is slow which may be due to the slow rate of degradation of lignin which contribute more to the char formation than the volatile.

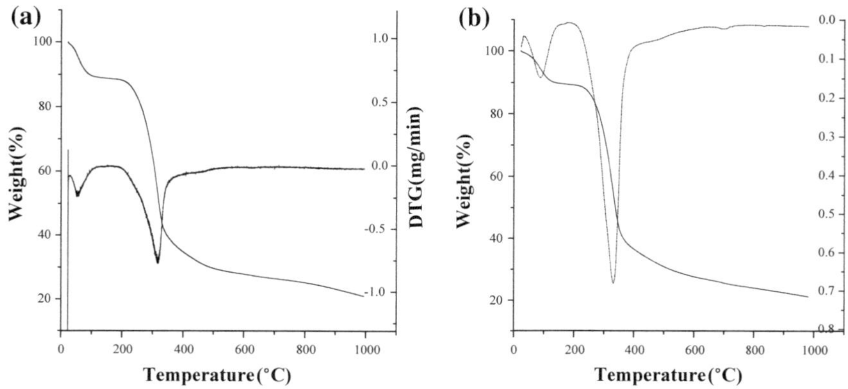

Fig. 7.2 TG-DTG curves of *Camellia sinensis* deoiled cake at heating rate of **a** 10 °C/min, **b** 40 °C/min

It can be observed from Table 7.2 that T_i, T_e, and T_{max} were laterally shifted to higher temperature values with increasing heating rate. This could be as at the same time and in the same temperature region, higher heating rate has a short reaction time and temperature needed for the sample to decompose is also higher which cause the maximum rate curve to shift right [7]. The analysis of the kinetic data was carried out assuming first order kinetics ($n = 1$). The kinetic parameter values calculated on the basis of the three methods are tabulated in Table 7.3. The non-uniformity of the rise and fall of activation energies in response to different heating rates applied makes sense to compare the results on an average basis (average of the results obtained at different heating rates) (Fig. 7.3).

The average activation energy of the active pyrolysis zone was calculated as 54.568 and 60.309 kJ mol^{-1} by Arrhenius and Coats–Redfern methods, respectively. The FWO method yielded 77.393 kJ mol^{-1} in the 0.1–0.9 conversion interval. In the active pyrolysis stage of CSDC, the results were almost closed for all of the methods applied for calculation. Table 7.3 shows excellent correlation with R^2 coefficient ranging from 0.9532 to 0.9806 over the temperature range used. Table 7.3 shows that the activation energy depends on the heating rate. As the slope of the straight line increase with increase in heating rate, the activation energy also increases. It may be due to the mineral presence in CSDC catalyze the cracking reaction which affect the values of activation energy.

Table 7.2 Temperature characteristics

Samples	Heating rate (°C/min)	Temperature (°C)		
		T_i	T_{max}	T_e
CSDC	10	118	316	400
	40	156	331	421

Table 7.3 Kinetic parameters calculated using different methods

Heating rate (°C/min)		E, kJ mol^{-1}	Log A, min^{-1}		Fitting plot equation				R^2
Arrhenius method									
10	52.278		4.409		$Y = -6288x + 7.8504$				0.9628
40	56.859		2.303		$Y = -6839x + 3.0221$				0.9532
Average	54.568		3.356						
Coast and Redfern method									
10	57.142		3.113		$Y = -6873x + 4.8648$				0.9806
40	63.477		1.457		$Y = -7635.6x + 1.0535$				0.9725
Average	60.309		2.284						
Activation energy calculated by FWO method (kJ mol^{-1})									
Conversion degree									
10	20	30	40	50	60	70	80	90	Average
93.698	114.118	92.951	109.789	97.672	101.996	94.472	100.225	100.225	77.393

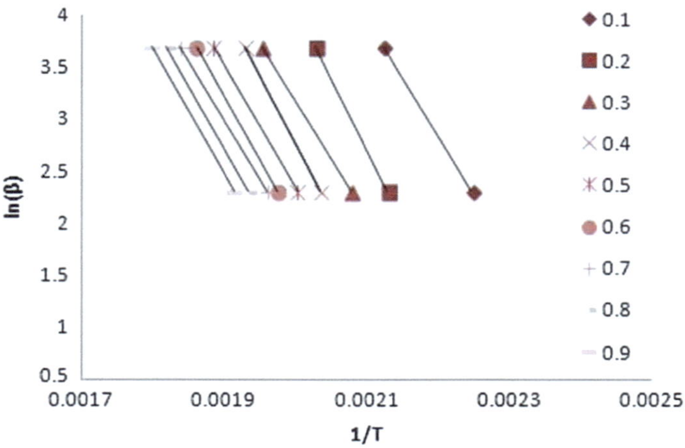

Fig. 7.3 Plot obtained by FWO method to calculate activation energy

7.5 Conclusion

In the present work, pyrolysis of CSDC has been carried out non-isothermally in 22–1000 °C temperature range at two different heating rates 10 and 40 °C/min under nitrogen atmosphere using thermogravimetric analyzer.

(1) It was observed that ~60% pyrolysis conversion can be obtained at a relatively low temperature (~400 °C). It indicates CSDC has a potential either to be converted to biofuel or use as fuel for energy purpose.
(2) The values of kinetic parameters of the active pyrolysis stage were calculated by using first order Coast-Redfern, Arrhenius, and model free FWO method. The values of activation energy E increased in proportion to heating rate. The activation energy of CSDC is in the range of 54–77 kJ mol^{-1}, this resembles other biomass materials which are in the range of 60–200 kJ mol^{-1} [8].

References

1. Gottipati R, Mishra S (2011) A kinetic study on pyrolysis and combustion characteristics of oil cakes: effect of cellulose and lignin content. J Fuel Chem Technol 39:265–270
2. Chutia RS, Phukan MM, Kataki R, Bhaskar T, Konwar BK (2016) Exploitation of Pongamia glabra deoiled cake for alternate energy: physico-chemical characterization and thermogravimetric studies. Energy Source Part A: Recovery Util Environ Eff 38:29–36
3. Vlaev LT, Markovska IG, Lyubchev LA (2003) Non-isothermal kinetics of pyrolysis of rice husk. Thermochim Acta 406:1–7
4. Hu S, Jess A, Xu M (2007) Kinetic study of Chinese biomass slow pyrolysis: comparison of different kinetic models. Fuel 86:2778–2788

5. Morris KW, Johnson WL (2000) In: 1st world conference on biomass for energy and industry, James and James (Science publishers), Sevilla, pp 1519–1524
6. McKendry P (2002) Energy production from biomass: overview of biomass. Bioresour Technol 83:37–46
7. Quan C, Li A, Gao N (2009) Thermogravimetric analysis and kinetic study on large particles of printed circuit board wastes. Waste Manag 29:2353–2360
8. Di Blasi C (2008) Modeling chemical and physical processes of wood and biomass pyrolysis. Prog Energy Combust Sci 34:47–90

Chapter 8
Development of Nanobased Thermic Fluid: Thermal Aspects of New Energy System

Shriram S. Sonawane and Vijay Juwar

Abstract In the present study, a cluster size of nanoparticle was evaluated using Xuan's model for water-based Fe_3O_4 nanofluid. The thermal conductivity of nanofluid with different nanoparticle concentration of 0.2, 0.4, 0.6 and 0.8 vol% was measured at temperature of 20, 40, 60 and 80 °C. It has been observed that thermal conductivity of nanofluid increases with an increase in concentration of nanoparticles and temperature of nanofluid. The agglomeration of nanoparticles in the fluid causes clustering, which adversely affects the thermal conductivity of nanofluid. To estimate the size of the cluster formed in the base fluid analytically, Xuan's model was employed. When experimental observations were compared with Maxwell's and Xuan's model, Xuan's model showed close proximity with observed results with an approximate cluster size of ten nanoparticles.

Keywords Nanofluid · Nanoparticles · Thermal conductivity

8.1 Introduction

Enormous consumption of fossil fuel is putting an increasing load of air, water and soil pollution on the earth. Unfavourable climatic changes are the results of green house effect, which is a major product of the air pollution. Stringent emission control norms can curb the air pollution but cannot minimize its level. As a part of solution to this problem, the research community across the globe gives a prime importance to the green and clean fuel. The reduction in consumption of fossils can drastically change the scenario so the researchers are looking forward to this issue in two ways, first is the utilization of alternative energy sources and later is to utilize fossils efficiently. To increase the efficiency of the thermal systems, nanofluid seems to be the promising option.

S.S. Sonawane (✉) · V. Juwar
Department of Chemical Engineering, Visvesvaraya National Institute of Technology (VNIT), Nagpur 440010, India
e-mail: shriramsonawane@gmail.com; sssonawane@che.vnit.ac.in

S. Kumar et al. (eds.), *Conference Proceedings of the Second International Conference on Recent Advances in Bioenergy Research*, Springer Proceedings in Energy, https://doi.org/10.1007/978-981-10-6107-3_8

Since inception nanofluid has shown its potential as new thermic fluid due to its enhanced thermal conductivity, when compared with conventional fluids like water, glycols, etc. In industrial sectors, including transportation, electronics, energy supply systems, heating and cooling fluids are of prime importance. For these fluids, thermal conductivity plays a pivotal role in deciding the rate of heating or cooling, requirements of quantity of fluid and size of energy exchange system. Nearly hundred years ago, Maxwell [1] showed that effective thermal conductivity of suspension containing spherical particles increases with an increase in the volume fraction of solid. Large surface area to the volume ratio is the basic requirement to enhance heat transfer through the fluid, because heat transfer takes place at the surface of the particle. By controlling shape and size of particle, Hamilton and Crosser [2] tried to evaluate its effect on heat transfer. Considering the effect of increased surface area to the volume ratio, the nanofluid came into picture. In comparison with suspensions of microsize particles, nanofluid had shown 1000 times larger surface area to the volume ratio, so the drastic improvement in thermal conductivity of nanofluid was observed. Maximum increment of 20% in thermal conductivity was observed by Eastmanan et al. [3] for nanofluid containing CuO and Al_2O_3 nanoparticle of size 35 nm suspended in water and ethylene glycol. Masuda et al. [4] observed that decrease in particle size, significantly increases the thermal conductivity of nanofluid. Sonawane et al. [5–12] observed increase in the volume fraction of nanoparticle increases the thermal conductivity of nanofluids. The nanoparticle suspended in the base fluid is in random motion and under the influence of several acting forces such as Brownian force and London-Van der wall forces. Aggregation and dispersion may occur among nanoparticles, clusters and individual nanoparticles during Brownian motion of the nanoparticle. The suspended nanoparticles may experience inter-particle collision and attachment of colliding particles and form aggregates under the influence of external and internal forces. Smaller size of colloid particles moves faster, so the energy transport inside the liquid becomes stronger. Furthermore, if the cluster grows large enough, it may sediment under the influence of gravity [13]. Thus, the formation of clusters inside the nanofluid adversely affects the energy transport within nanofluid. In the present work, an attempt is made to the study of Fe_3O_4 ethylene glycol-based nanofluid for the effect of temperature, nanoparticle volume fraction cluster formation on the thermal conductivity of nanofluid.

Increase in nanoparticle concentration enhances thermal conductivity and viscosity of nanofluid [14]. At higher concentration, nanoparticles tend to agglomerate due to its high surface energy. Due to agglomeration, sedimentation of nanoparticle takes place, which adversely affects the thermal conductivity of nanofluid [15]. Increase in viscosity due to agglomeration results in high pressure drop and increase in pumping cost which in turn decreases the heat transfer rate [15, 16]. To obtain uniformly dispersed nanofluid and to break up cluster formation, ultrasonication treatment is applied to nanofluids. Kole and Dey [17] reported enhancement of thermal conductivity of nanofluid with an increase in sonication time and maximum value obtained at 60 h of sonication time for ZnO ethylene glycol nanofluid. Yang et al. [18] observed that viscosity of CNT oil-based nanofluid kept decreasing with increasing ultrasonication time. Mahbubul et al. [16] studied alumina–water nanofluid and reported that viscosity of nanofluid decreases with temperature and extension of ultrasonication treatment; at lower temperatures, extended ultrasonication time was required to reach to lowest viscosity when compared with higher temperature. Thermal conductivity of nanofluid increases with increasing ultrasonication exposure. The thermal conductivity and viscosity of nanofluid were found to be increasing with an increase in nanoparticle concentration. However, temperature and ultrasonication exposure showed negative effects on viscosity and positive effect thermal conductivity [16]. For an ideal heat transfer media, minimum viscosity and maximum thermal conductivity are fundamental requirements. The condition of maximum thermal conductivity and minimum viscosity can be meeting through multi-response optimization of independent variables.

8.2 Material and Method

8.2.1 Nanoparticle Synthesis

Fe_3O_4 nanoparticles were synthesized using chemical precipitation method. Ferric chloride, ferrous chloride and sodium hydroxide were chemical precipitation method. All the chemicals used were of Fisher Scientific of SQ grade. The molar ratio of 2:1 of ferric chloride to ferrous chloride was used in its dissolution in distilled water. Once the precursors were fully dissolved, water became orange in colour. To maintain the pH of solution at 12, NaOH solutions were added drop by drop. A black colour precipitate was observed after 45 min. of vigorous stirring. The precipitate was washed several times with acetone and distilled water and dried at 80 °C for 24 h [19]. With the help of atomic force microscopy (AFM), the size of nanoparticle was estimated and found to be 34 nm, as shown in Fig. 8.1.

8.2.2 Experimental Set up

The Fe_3O_4 nanoparticle of average size 34 nm and density 5.16 gm/cc was dispersed in ethylene glycol. For uniform dispersion of nanoparticles in base fluid, ultrasonication treatment was given for 45 min. While making samples, high precision mass balance of Shimadzu was used for weighing nanoparticles. Nanofluids, with different volume concentrations of 0.2, 0.4, 0.6 and 0.8%, were dispersed in ethylene glycol. A ChromTech sonicator (Taiwan) with 40 kHz and 1200 W was used for ultrasonication treatment. The thermal conductivity of nanofluid was measured using KD-2 PRO thermal property analyzer (Decagon Devices, USA). It works on the principle of transient hot wire method (THW) [15]. The THW system used in KD2 system infers thermal conductivity from the temperature response of a

Fig. 8.1 AFM image of Fe_3O_4 nanoparticles

thermocouple a short distance away from an electrically heated wire. The uncertainty of the KD2-pro for thermal conductivity measurement as indicated by the manufacturer is ±5%. The thermal conductivity of every sample was measured at four different temperatures 20, 40, 60, and 80 °C, respectively.

8.2.3 Modelling of Thermal Conductivity

After experimentation, it becomes necessary to compare the experimental results with theoretical models. Some of the widely used models are given below.

Maxwell [1] proposed the model for thermal conductivity of solid–liquid solution as

$$k_{\mathrm{eff,\,Maxwell}} = \frac{k_p + 2k_b + 2\left[k_p - k_b\right]ø}{k_p + 2k_b - \left[k_p - k_b\right]ø} k_b \tag{8.1}$$

where k_p is the thermal conductivity of particle, k_b thermal conductivity of base fluid and ø is the volume fraction.

In 1962, Hamilton and Crosser [2] proposed a model for liquid–solid mixtures of non-spherical particles. They introduced a shape factor, 'n', to account for the effect of the shape of the particles. The thermal conductivity, in which the ratio of conductivity of the solid and fluid phases is larger, than 100 (k_p, $k_b > 100$), can be expressed as follows:

$$k_{\mathrm{eff,\,Hamilton}} = \frac{k_{\mathrm{p}} + (n-1)k_{\mathrm{b}} - (n-1)\left[k_{\mathrm{b}} - k_{\mathrm{p}}\right]ø}{k_{\mathrm{p}} + (n-1)k_{\mathrm{b}} + \left[k_{\mathrm{b}} - k_{\mathrm{p}}\right]ø} k_{\mathrm{b}} \tag{8.2}$$

where 'n' is the empirical shape factor given by $n = 3/\psi$, and 'ψ' is the particle sphericity defined as the ratio of the surface area of a sphere with volume equal to that of the particle to the surface area of the particle. The classical models are derived from continuum formulations and include only the particle shape and volume fraction as variables and assumed diffusive heat transport in both liquid and solid phases. The large enhancement of the effective thermal conductivity in nanofluids challenges Maxwell's theory [1] as well as its modification by Hamilton and Crosser [2]. Some important mechanisms in nanofluids appear to be neglected in these models.

Koo and Kleinstreuer [20, 21] developed a new model for nanofluids, which includes the effects of particle size, particle volume fraction and temperature dependence as well as properties of the base fluid [k_{Static}] and the particle subject to Brownian motion [k_{Brownian}].

$$k_{\mathrm{eff}} = k_{\mathrm{static}} + k_{\mathrm{Brownian}}$$

The resulting formula is

$$k_{\mathrm{eff,\,Koo}} = \frac{k_{\mathrm{p}} + 2k_{\mathrm{b}} + 2\left[k_{\mathrm{p}} - k_{\mathrm{b}}\right]ø}{k_{\mathrm{p}} + 2k_{\mathrm{b}} - \left[k_{\mathrm{p}} - k_{\mathrm{b}}\right]ø} k_{\mathrm{b}} + 5 \times 10^{4} \beta ø \rho_{\mathrm{p}} c_{\mathrm{p}} \sqrt{\frac{k_{\mathrm{b}} T}{\rho_{\mathrm{p}} D}} f[T, ø] \tag{8.3}$$

Xuan et al. [22] introduced the model on the basis of the theory of Brownian motion, and the diffusion-limited aggregation model is applied to simulate random motion and the process of aggregation of the nanoparticles. A theoretical model is developed to predict the thermal conductivity of nanofluids.

$$\frac{k_{\mathrm{eff}}}{k_{\mathrm{f}}} = \frac{k_{\mathrm{p}} + 2k_{\mathrm{f}} - 2ø\left(k_{\mathrm{f}} - k_{\mathrm{p}}\right)}{k_{\mathrm{p}} + 2k_{\mathrm{f}} + ø\left(k_{\mathrm{f}} - k_{\mathrm{p}}\right)} + \frac{\rho_{\mathrm{p}} ø c_{\mathrm{p}}}{2k_{\mathrm{f}}} \sqrt{\frac{k_{\mathrm{B}} T}{3\pi\eta r_{\mathrm{c}}}} \tag{8.4}$$

where k_{B} is Boltzmans's constant 1.3810×10^{23} J/K, ρ_{p} is the density of particle and r_{c} is the cluster radius, η dynamic viscosity.

In the present work, Maxwell's model and Xuan's model were used to compare with the experimental results.

8.3 Results and Discussion

The effect of increasing temperature on the thermal conductivity of nanofluid is as shown in Fig. 8.2.

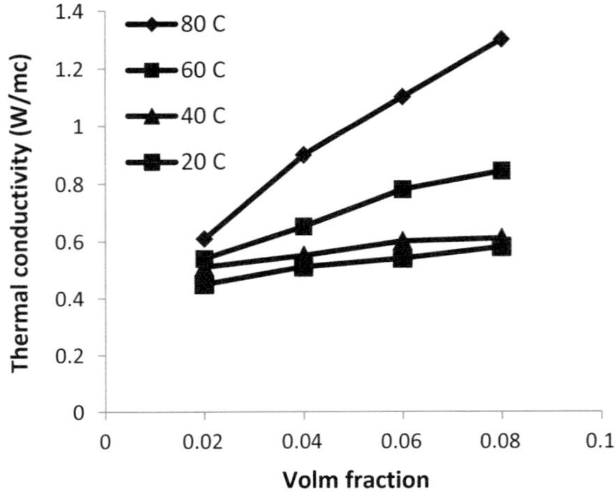

Fig. 8.2 Effect of increasing temperature on thermal conductivity of nanofluids

As shown in Fig. 8.2, as the temperature of nanofluid increases, the thermal conductivity of nanofluid increases. Due to increase in temperature, the Brownian motion of nanoparticle increases which in turn increases thermal transport within the fluid, this may be the probable reason behind increment of thermal conductivity with temperature. From figure, it is very much clear that the thermal conductivity of nanofluid increases with increase in volume fraction of nanoparticles in nanofluid. With increase in volume fraction of nanoparticle in nanofluid, the inter-particle distances within fluid decreases which are helpful in energy transport cross the fluid which in turn increases the thermal conductivity of nanofluid. The number of experiments is repeated about three times, and average results are used for this figure.

As shown Fig. 8.3a–d, experimental results are in close proximity of Xuan's model, whereas Maxwell's model underestimates the thermal conductivity of nanofluid. In Maxwell's model, effect of temperature of nanofluid is not taken into consideration as well as the size of nanoparticle in the fluid is not considered in the estimation of thermal conductivity of nanofluid; however, in Xuan's model, the diffusivity and Brownian motion of nanoparticle are taken as the basis of estimation of thermal conductivity. The nanoparticles are having very high surface area to volume ratio, so there is a tendency of clustering of nanoparticles in the fluid. Clusters so formed in the fluid adversely affect the energy transport across the fluid. Increase in the size of cluster starts the sedimentation of clusters under the influence of gravity and reduces the energy transport which in turn decreases the thermal conductivity of nanofluid. Increase in temperature reduces the inter-particle and intra-particle forces in the nanofluid which decrease the clustering tendency of nanoparticles in nanofluid. The term r_c used in Eq. 8.4 is the radius of cluster formed in the fluid. When no clusters form, the radius of nanoparticle is equal to radius of cluster. In the present study, Xuan's model comes in close proximity of

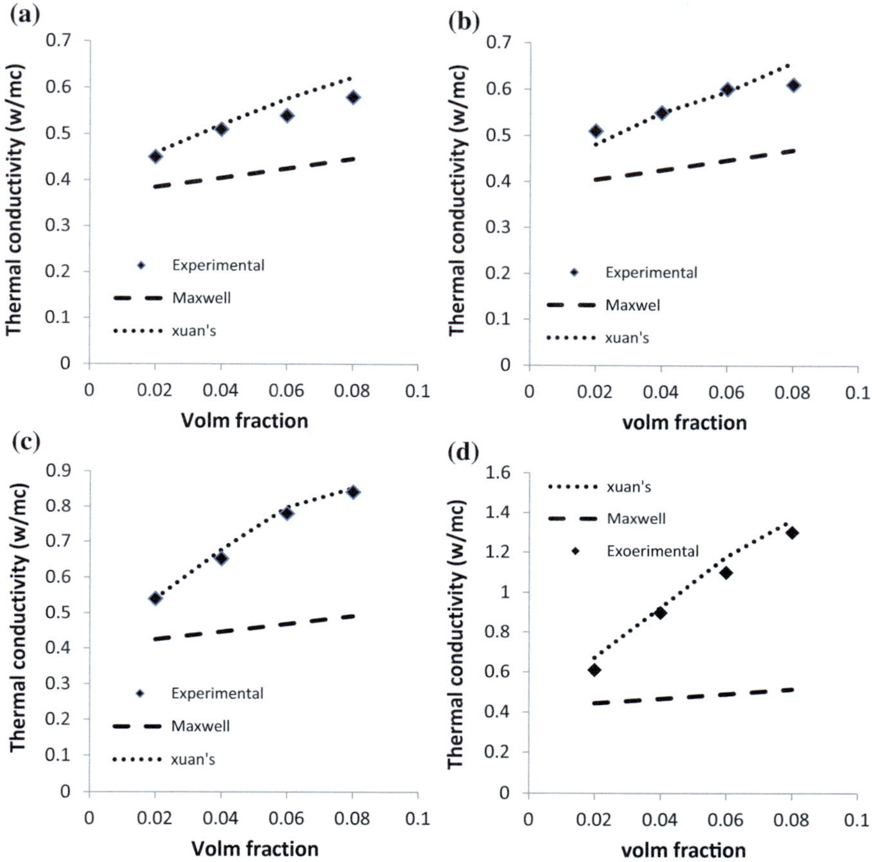

Fig. 8.3 Comparison of experimental results with Maxwell's and Xuan's model. **a** 20 °C, **b** 40 °C, **c** 60 °C, **d** 80 °C

experimental results and radius of cluster is nearly equal to ten times the radius of nanoparticle.

8.4 Conclusions

The thermal conductivity of nanofluid was found to be increasing with increasing temperature and volume fraction of nanoparticles in nanofluid. Maxwell's model underestimates the thermal conductivity of nanofluid within experimental range. The experimental results were found to be in close proximity with Xuan's model in the experimental range. By Xuan's model, approximate size of cluster formed in the nanofluid was found to be equal to ten times the nanoparticle radius.

References

1. Maxwell JC (1904) A treatise on electricity and magnetism, 2nd edn. Oxford University Press, Cambridge, pp 435–441
2. Hamilton RL, Crosser OK (1962) I&EC Fundam 1:187
3. Eastmanan JA, Choi SUS, Li S, Yu W, Thompson LJ (2001) Appl Phys Lett 78(6)
4. Masuda H, Ebata A, Teramae K, Hishinuma N (1993) Netsu Bussei 4:227
5. Khedkar RS, Sonawane SS, Wasewar KL (2013) Synthesis of TiO_2–water nanofluids for its viscosity and dispersion stability study. J Nano Res 24:26–33
6. Sonawane SS, Khedkar RS, Wasewar KL (2014) Study on concentric tube heat exchanger heat transfer performance using Al_2O_3–water based nanofluids. Int Commun Heat Mass Transf 49:60–68
7. Khedkar RS, Sonawane SS, Wasewar KL (2014) Water to nanofluids heat transfer in concentric tube heat exchanger: experimental study. Procedia Eng 51:318–323
8. Khedkar RS, Sonawane SS, Wasewar KL (2014) Heat transfer study on concentric tube heat exchanger using TiO_2–water based nanofluid. Int Commun Heat Mass Transf 57:163–169
9. Khedkar RS, Kiran AS, Sonawane SS, Wasewar K, Umre SS (2013) Thermo-physical characterization of paraffin based Fe_3O_4 nanofluids. Procedia Eng 51:342–346
10. Sonawane SS, Juwar V (2013) Investigations on rheological behaviour of paraffin based Fe_3O_4 nanofluids and its modelling. Res J Chem Environ 19(12):22–29
11. Juwar V, Sonawane S (2013) Development of nano based thermic fluid: rheological aspects of new energy system. Recent Adv Bioenergy Res 433
12. Khedkar RS, Sonawane SS, Wasewar KL (2014) Synthesis of TiO_2–water nanofluids for its viscosity and dispersion stability study. J Nano Res 24:26–33
13. Gregory, Fendler JH, Dekany I (eds) (1996) Nanoparticles in solid and solutions. Kluwer, Boston, p 203
14. Li Y, Zhou JE, Tung S, Schneider E, Xi S (2009) A review on development of nanofluid preparation and characterization. Powder Technol 196(2):89–101
15. Ruan B, Jacobi AM (2012) Ultrasonication effects on thermal and rheological properties of carbon nanotube suspensions. Nanoscale Res Lett 7(1):127
16. Mahbubul IM, Shahrul IM, Khaleduzzaman SS, Saidur R, Amalina MA, Turgut A (2015) Experimental investigation on effect of ultrasonication duration on colloidal dispersion and thermophysical properties of alumina–water nanofluid. Int J Heat Mass Transf 88:73–81
17. Kole M, Dey TK (2012) Effect of prolonged ultrasonication on the thermal conductivity of ZnO–ethylene glycol nanofluids. Thermochim Acta 535:58
18. Yang Y, Grulke EA, Zhang ZG, Wu G (2006) Thermal and rheological properties of carbon nanotube-in-oil dispersions. J Appl Phys 99:114307
19. Healy JJ, de Groot JJ, Kerstin J (1976) Physica B+C 82:392–408
20. Koo J, Kleinstreuer C (2005) A new thermal conductivity model for nanofluids. J Nanopart Res 6(6):577–588
21. Koo J, Kleinstreuer C (2005) Impact analysis of nanoparticle motion mechanisms on the thermal conductivity of nanofluids. Int Commun Heat Mass Transf 32(9):1111–1118
22. Xuan Y, Li Q, Hu W (2003) Thermodynamics. AIChE J 49(4)

Chapter 9
Strategies for Producer Gas Cleaning in Biomass Gasification: A Review

Haresh V. Makwana, Darshit S. Upadhyay, Jayesh J. Barve and Rajesh N. Patel

Abstract Bioenergy is the prominent source of generation of power by means of thermal and electricity route. The potential of biomass, a renewable energy source, is estimated to be 500 million metric tons per year in India. Gasification, a promising route to extract energy from biomass, is a thermochemical process having high thermal efficiency compared to direct combustion. Producer gas is the outcome of gasification. Producer gas is a medium calorific value gas and is rich in H_2 and CO. Despite innumerable applications of the biomass gasification gas, this century-old technology is yet not in its mature form. This is because of impurities like tar and particulate matter (PM). Tar is condensable at the downstream of gasifier system and leads to clogging or blockage in the process equipment. PM can create problems like fouling and corrosion. In this review paper, various methodologies are discussed which can diminish or destruct tar and PM. The purpose of this review is to analyze the recent technologies to clean the producer gas. Various gas cleaning methods like physical, non-catalytic, and catalytic are also discussed. The findings of the review are that catalytic tar conversion has the potential to increase conversion efficiencies while simultaneously decreasing tar, and therefore it is an instinctively effective method to obtain clean gas.

Keywords Gasification · Tar · Particulate matter (PM) · Gas cleaning

H.V. Makwana · D.S. Upadhyay (✉) · R.N. Patel
Mechanical Engineering Department, Institute of Technology, Nirma University, Ahmedabad 382481, Gujarat, India
e-mail: darshitupadhyay@yahoo.com; darshit.upadhyay@nirmauni.ac.in

J.J. Barve
GE Global Research, GE India Technology Center, Bangalore, India

S. Kumar et al. (eds.), *Conference Proceedings of the Second International Conference on Recent Advances in Bioenergy Research*, Springer Proceedings in Energy, https://doi.org/10.1007/978-981-10-6107-3_9

9.1 Introduction

9.1.1 Need of Alternative Energy Sources

It is well-known that majority of power (heat and electricity both) is derived from the conventional energy sources such as fossil fuels. The world is facing a major problem with greenhouse gases emission like CO_2 by burning fossil fuels which cause global climate change. With depletion of these conventional fuels as well as rising costs, the focus of worldwide researchers and scientists is mainly on alternative source for reducing dependency on conventional fuels as world's energy supply. Renewable energy has been around for some time, and hence, a matter of discussion as alternative energy sources. Renewable energy sources like solar, wind, biomass, and small hydro are considered as having good potential to meet the huge power demand over the globe to some extent. It is estimated that new investments in renewable power and fuels have increased from 2004 to 2013 abruptly as shown below in Table 9.1 [1].

Table 9.1 Renewable energy indicators 2013 [1]

		Start 2004	End 2012	End 2013
Investment				
Annual new investment in renewable energy	Billion USD	39.5	249.5	214.4
Power				
Renewable power capacity (not considering hydro)	GW	85	480	560
Renewable power capacity (including hydro)	GW	800	1440	1560
Hydro power capacity	GW	715	960	1000
Bio power capacity	GW	<36	83	88
Bio power generation	TWh	227	350	405
Geothermal power capacity	GW	8.9	11.5	12
Solar PV capacity	GW	2.6	100	139
Concentrating solar thermal power	GW	0.4	2.5	3.4
Wind power capacity	GW	48	283	318
Heat				
Solar hot water capacity	GW	98	282	326
Transport				
Ethanol production	Billon liters	28.5	82.6	87.2
Biodiesel projection	Billon liters	2.4	23.8	26.3

9.1.2 Bioenergy

Biomass is one of most trusted energy sources among all renewable sources. Total installed capacity has increased from 41 GW in 2005 to 88 GW in 2013. Estimated growth rate of biomass power usage is 12% in 2013 which is significantly higher than 4% in 2005 [2]. In India, estimated potential of biomass power is 17.538 GW and its share in renewable energy power generation amounts to 11.88% as of 2014 [3].

As one of the largest sustainable energy sources worldwide, biomass is widely used for thermal energy, electric power, and other chemical by-products (residue). Biomass, as per its inherent nature, circulates carbon dioxide (CO_2) and is helpful in controlling global climate change. It consists of organic materials produced by photosynthesis process in the presence of sunlight. It absorbs solar energy and stores it in its chemical bonds as a chemical energy [4], which is further released/used to break down the bonds [5]. The energy stored in biomass can be released/utilized via thermochemical conversion and biological/biochemical conversion. In thermochemical conversion, chemical bonds have to be broken for releasing stored chemical energy. In processes like direct combustion, gasification, liquefaction, and pyrolysis, heat is involved. For releasing energy, direct combustion process involves igniting biomass in the presence of air/oxygen, whereas in other processes, energy is transferred into the gas/liquid as a secondary product. In biological/biochemical conversion, bacteria, biocatalyst or other microorganism are used to break down biomass to liquid fuels. In process like fermentation and anaerobic digestion, heat is not at all involved. Based on the advantages in terms of energy efficiency and ease of application, gasification is relatively the best choice for producing energy from biomass [6].

9.1.3 Gasification to Produce Biomass

Gasification is conversion of biomass into producer gas which has lower/medium calorific value by using gasifying agents such as oxygen, steam, and air. The product gas is called producer gas or synthesis gas which is rich in hydrogen and carbon monoxide. This gas is directly ignited in various devices, dryers, kilns, furnaces, and boilers in industries e.g., ceramic industries and rubber industries, for thermal purpose because of fuel prices being lower than natural gas. Since the gas is rich in H_2 and CO, it can be separated to utilize for fuel cell [7, 8] or to convert into liquid hydrocarbon fuels or chemicals by the Fischer–Tropsch synthesis method [9, 10]. Also, this gas can be used in reciprocating engine such as I.C. Engine or diesel generator and rotary gas turbine for power generation [11]. As compared to direct combustion of biomass, gasifying biomass into producer gas helps not only in improving thermal efficiency but also in limiting air pollution from the conversion process. Gasification is a century-old technology, which gained prominence during World War II (WWII) especially in Germany to reduce dependency on other countries for liquid fuels. After WWII, crude oil was available at cheaper costs from many regions of the world especially from the Middle East shifting to large-scale usage of petroleum. Because of the same, gasification technology did not receive much attention comparatively and still a lot of untapped scope of development exists with respect to this technology.

Different types of gasifiers are available in market such as fixed bed/moving bed, fluidized bed, and entrained flow. Yet, gasifiers are not widely popular especially for power generation purpose. Along with low/medium calorific value producer gas, other by-products are also emitted. Some of these by-products, i.e., tar, particulate matter, NH_3, H_2S, NO_x, HCl, SO_2 are difficult to handle and create numerous problems in the downstream of the gasifier system [12–16].

9.1.4 By-Products: Tar and Particulate Matter

Various researchers have defined tar in their own way. In gasification, tar is operationally defined as a condensable material in downstream processing steps or conversion devices [17]. Brownish, typical smelling, highly viscous, and sticky in nature, tar can create problems such as cracking in the pores of filters, block the process equipment like pipe, nozzles and valve, forming coke and plugging them, condensing in the cold spots and plugging them, and many more. High concentrations of tars can damage or lead to unacceptable levels of maintenance for engines and turbines.

Particulate matter is defined as particles suspended in the gas in the form of minute solid particles or liquid droplets. It can create problems like agglomeration and slagging, deposition, fouling, and corrosion. Pollutants can be removed in the gasifier reactor itself or in the downstream of the system.

Thus, it can be inferred from the above-mentioned facts that cleaning of producer gas is a very important and essential aspect. There are major three techniques to diminish pollutant from producer gas: (1) Physical Method, (2) Thermal Cracking, and (3) Catalytic Cracking. In physical method, various physical systems have been used such as various filters, scrubbers, separators, and similar systems. This method is convectional method and lots of research has been done. Pressure drop at each system is the major issue with this system. In thermal cracking process, producer gas is heated at higher temperature. For the same, small portion of producer gas has been used for heating purpose. This method has the ability to provide almost tar and PM-free gas. Gas production is reduced which leads less thermal efficiency is a drawback of this method. In catalytic cracking, various catalysts have been used in gasifier system. Selection of catalyst is mainly dependent on the type of gasifier and feedstocks used. This method is preferred over all methods because it has the ability to clean producer gas meanwhile it can increase the conversion efficiency.

There are two types to reduce/remove pollutant from producer gas: (1) Primary Method and (2) Secondary Method. In Primary method, pollutant is removed at gasifier reactor itself. In secondary method, pollutant is removed at the downstream of the gasifier reactor. In general, study of hot gas cleaning techniques and cold gas cleaning techniques are important. Lots of research had been done in hot gas cleaning techniques. Most of the thermal and catalytic cracking methods fall in this category. Wide numbers of catalyst can be able to react at high temperatures. For thermal cracking also, high amount of heat is required. These techniques are preferred as it cannot restrict the flow like mechanical system. Some applications of both the techniques are described in this paper.

In this review paper, various cleaning methods are described with relevant advantages and disadvantages. Cleaning of producer gas through physical, thermal cracking, use of catalyst, and steam gasification have been reviewed.

9.2 Tar and PM Removal by Physical Method

Tar formation is due to incomplete conversion of fuel into gas in gasification. This happens due to various reasons like the physical design of gasifier and chemical reaction in the process

Devi et al. [18] reported that electrostatic precipitators (ESPs) can capture particulate matter effectively from the producer gas. The efficiency of ESPs is high for fine-mode PM. ESPs can handle very high temperatures of up to 700–800 °C with collection efficiencies up to 99% for particles smaller than 10 nm. An electrostatic precipitator (ESP) is a filtration device that removes fine particles, like dust and smoke, from a flowing gas using the force of an induced electrostatic charge minimally impeding the flow of gases through the unit.

Machin et al. [19] studied that changing the configuration in the combustion chamber of gasifier (i.e., primary method) reduces tar up to 10 mg/Nm^3 and is used for the automobile (Otto) engine. Several mechanisms such as filter, cyclone, and ESP are deployed to trap particles. Tar removal efficiency from 51 to 91% is accounted for venturi scrubber to purge the gases in a countercurrent for rice husk gasifier [20]. Bridgwater [21] found that tar in the fuel gases was lower than the 20–40 mg/Nm^3 for high productivity of venturi scrubber.

Using the wet scrubbing method, PM can be removed. Venturi scrubber works on the principle of increasing gas velocity by reducing the flow area, spraying water into fine droplets. The efficiency is more than 50% [22]. Tar can be reduced by primary and secondary methods. Primary method is to change various parameters in the gasifier itself such as ER, retention time, but this is not suited for the gas engine application.

By corona plasma, tar can be eliminated at higher temperature [23]. Cyclone separator is used to separate the syngas from the dirt and ash entrained within the gas stream. The basic principle behind cyclone separators is deploying force to separate dirt particles from the gas stream. A cone stream causes the vortex diameter to decrease till the gas reverses on itself and spin up from the middle to the outlet pipe or vortex finder. This method has 55% efficiency [22].

Table 9.2 shows that the greater amount of quality gas can be obtained by the high retention time. By using the floating drum gas storage system, the content of tar produced in the producer gas after 3.5 h of retention time was 9.24 mg/Nm^3.

Chiranjeevaraoseela and Vykunta [24] developed a tar-measuring technique that uses different adsorbent on the bed of combustion zone, activated carbon shows the best performance reducing the content of tar from 24 to 4 g.

Various gas cleaning techniques are shown in Fig. 9.1. They are classified based on wet scrubbing, dry or wet scrubbing, and hot gas cleaning. The acceptable range of tar, PM, and particle size for reciprocating I.C. engine and rotary gas turbine is displayed (Table 9.3).

Candle filter is built of dirt trap in which material silicon can be carbide (SiC) and other composite ceramics. It can operate at temperatures up to 500 °C and can adequately uproot particles in the 0.5–100 μm range. The fired flame channel contains a nickel-based tar splitting impetus in the bolster body. Tar evacuation proficiency is somewhere between 96 and 98% for naphthalene and 41–79% for benzene [22] (Table 9.4).

Ma et al. [26] reported that two stage gasification where secondary air was supplied to the combustion zone found better as it can easily crack the tar in producer gas. The average value of gas generated was 4.7–5.28 MJ/Nm^3. Hindsgaul [27] studied that cleaning biomass through the bag and cartridge filter has good cleaning efficiency (96–99% by mass). It is an effective technique for particle removal from syngas with low tar concentration.

Neubauer [28] proposed that there is no single parameter or technique involved to reduce tar in biomass gasification. It also depends on the plant size from kW to MW range as per the feedstock. Hofbauer et al. [29] observed that by using natural catalyst as a bed material, temperature above 750 °C in the FICFB (fast internally circulating fluidized bed)—tar content was reduced below 3 g/Nm^3.

Han and Kim had done critical comparison between various mechanical systems along with their particle removal and tar removal efficiencies.

Table 9.2 Tar content of producer gas at different retention time (Yang wood, MC 5 8.3% $_{\text{wet-basis}}$) [22]

Primary and secondary air flow rate (L/min)	Retention time (min)	Tar content (mg/Nm^3)
80	16.5	40.58
80	21.0	22.74
80	28.7	22.22
80	204.7	9.24

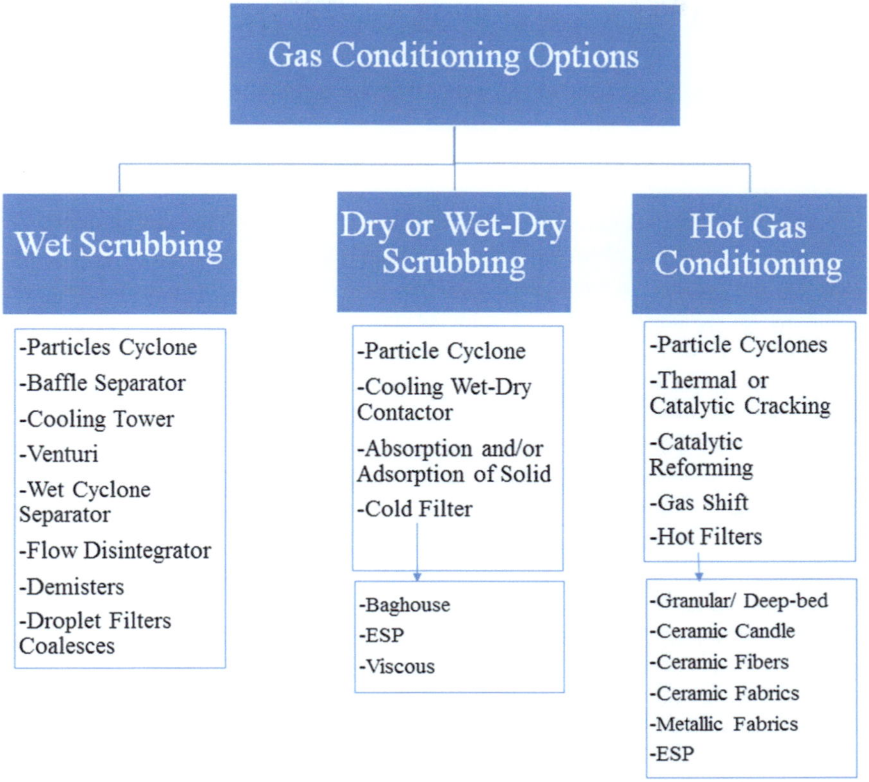

Fig. 9.1 Gas cleaning techniques [24]

Table 9.3 Gas quality requirements for power generation [25]

Constituents	Units	I.C. engine	Gas turbine
Particles	mg/Nm^3	<50	<30
Particle size	μm	<100	<5
Tar	mg/Nm^3	<10	N/A

Table 9.4 Tar and particle removal efficiency of some mechanism methods [30]

Mechanical methods	Particle removal (%)	Tar removal (%)
Sand bed filter	70–99	50–97
Wash tower	60–98	10–25
Venturi scrubber	N/A	50–90
Wet electrostatic precipitator	>99	0–60
Fabric filter	70–95	0–50
Rotational particle separator	85–90	30–70
Fixed bed tar adsorber	N/A	50

9.3 Tar and PM Removal by Thermal Cracking

Compounds of tar are shown in Table 9.5. PM can be classified as $PM_{2.5}$ small size which is fine particles containing more tar and less char, whereas PM_{10} big size is coarse particles containing less tar and more char. It is effective technique for particle removal from syngas with low tar concentration. Laurence and Ashenafi [33] studied the various parameters during the gasification of biomass can influence the content of tar. They observed that the tar yield decreased from 15 to 0.54 g/Nm^3 when the temperature is increased from 697 to 817 °C. As the ER increases, the amount of tar decreases with heat value of gas.

A design of 20 kW using wood is found to be an effective method to clean gas after the gas passes through filter with maximum retention time. The content of tar can be reduced from 24 to 53.52 mg/m^3 using filter [34]. Nakamura et al. [35] reported that use of bio-oil scrubber 100 L with char bed 10 and 3 kg of activated carbon can remove tar about 60% from the producer gas. The problem of performance decreasing to 40% after 10 h of operation is because the surface of char bed gets covered by the tar with adsorption of char.

Asadullah [36] has investigated the biomass gasification for gas quality and use of catalyst to clean the producer gas. Hot gas cleaning is more effective than cold gas. Catalytic technique results into less tar and improves gas quality. The fouling of IC engine and its pipelines is caused by tar class 5, 4, and 2. Several researchers have challenged the treatment of producer gas used for IC engine application. Thermal and catalytic methods are more attractive due to complete destruction of

Table 9.5 Basic approximations for tar compound classification [31, 32]

Class	Description	Properties	Compounds
1	GC-undetectable	Very heavy tar cannot be detected by GC	Determined by subtracting the GC-detectable tars from the total gravimetric tar
2	Heterocyclic aromatics	Tars containing hetero atoms; highly water soluble	Pyridine, phenol, quinoline, isoquinoline, dibenzophenol, cresols
3	Light aromatic (1 ring)	Usually single ring light hydrocarbons; do not pose a problem regarding condensation or solubility	Toluene, ethylbenzene, xylenes, styrene
4	Light PAH compounds (2–3 rings)	2 and 3 ring compounds; condensation occurs at low temperatures even with low concentrations	Indene, naphthalene, methylnapthalene, biphenyl, acenaphthalene, fluorene, phenanthrene, anthracene
5	Heavy PAH compounds (4–7 rings)	Larger than 3 ring; condensation occurs at high temperatures even with low concentrations	Fluoranthene, pyrene, chrysene, perylene, coronene

tar and attaining higher efficiency [37]. It is equipped with burner at the lower part of the pyrolysis zone for reducing the amount of water content in the gas and increasing the gas temperature. There is no tar release out of reactor due to direct combustion. LHV of the producer gas was observed to be 5,012 kJ/Nm^3 [38].

(PID) Photo ionization detector is the fastest online method to detect the amount of individual compounds in the producer gas such as naphthalene, anthracene, and fluorene Anthracene, pyrene, and other aromatic compounds can be found at temperature 850 °C by gas chromatography [39]. Akudo and Theegala [40] investigate the effective technique to get higher heating value from the syngas via sampling train. The average value of tar and particulate concentration ranges from 1.8 to 3.1 and 5.2 to 6.4 g/m^3. The heating value of producer gas ranges from 4.38 to 4.55 MJ/m^3. Ample numbers of tar and PM measuring techniques were developed by various researchers.

9.4 Tar Reduced by Steam Gasification

Ligang et al. [41] observed that the presence of steam increases the gas yield and reduces the tar and char yield in downdraft gasifier. With the increase of temperature (750–850 °C), the tar content of the gas from legume straw decreases from 62.8 to 3.7 g/Nm^3 dry gas, while that from pine sawdust decreases from 45.6 to 6.0 g/Nm^3 dry gas. It can be found that an increase in H_2 and a decrease in CO concentration of product gas take place at elevated temperature. (H_2 concentration of 50.6 and 44.0 mol% and CO concentration of 21.2 and 28.2 mol% from legume straw and pine sawdust, respectively, at 850 °C). Franco et al. [42] carried out work on fluidized bed gasifier by taking gasifier agent as steam. They observed if temperature is increased from 700 to 900 °C, then H_2 increases 10–20% (26–33 mol%) and CO slightly decreases (41–38 mol%). Also, they optimized operating parameters like steam/biomass ratio 0.6–0.7 w/w at 830 °C leading to higher H_2%, carbon conversion, greater gas yield and poorer hydrocarbon and tar. Saw and Pang [43] had done experiments using duel fluidized bed steam gasifier (100 kW) with blended lignite and wood pellets. They observed lignite to wood ratio increasing from 0 to 100%, the H_2/CO ratio increased asymptotically from 1.0 to 2.4 while the tar concentration and tar yield decreased from 9.0 to 2.7 g/Nm^3 and from 6.6 to 2.3 g/kg of dry fuel, respectively. Aitziber et al. [44] carried out work on conical-spouted bed reactor taking pinewood as a fuel. They observed increasing temperature leads to increase in H_2, decrease in CO and tar, and decrease in char yield concentration. They also noticed steam/biomass increase (0–1) gives positive effect on hydrogen concentration, char gasification, and tar reforming but steam/biomass (1–2) gives a limited improvement in the results.

9.5 Tar Reduced by Catalytic Cracking

It was found that Canadian limonite iron to be more active than olivine as the bed material for tar reduction in fluidized bed gasification of biomass at a small equivalence ratio. During experiment, very low tar yield was observed of 15–25 g/kg of biomass at equivalence ratio 0.3 [45].

Dong et al. [46] had experimentally found tar concentration in producer gas up to 82 mg Nm^{-3} and the tar conversion was up to 97% with coal char supported iron catalyst in updraft gasifier. Study showed that catalyst's surface area and pore volume were not the dominant factors affecting the activity of catalysts as well as particle size of catalyst had minor effects on gas compositions [47]. K-based catalyst with steam as gasifying agent showed a significant effect on efficiency [48]. With NiO on modified dolomite (NiO/MD) catalyst, tar is mostly eliminated while gas and hydrogen yield increase (50.49%) significantly [49]. CaO increases the formation of H_2 by interacting with producer gas and catalyzing thermal cracking of tar to some extent [50]. By using natural catalyst and temperature more than 750 °C, tar can be reduced to below 3 g/Nm^3. Moreover, tar reduction efficiency was approximately 50% at 600 °C gasification temperature and 20% tested catalyst addition [51, 52]. With use of catalyst and steam gasification, tar concentration is reduced abruptly and H_2 yield is increased in product gas [53–55]. As catalyst granular size increases, H_2 in syngas increases from 30.52 to 34.22% and CO decreases from 32.31 to 28.86% at optimum condition [56].

9.6 Conclusion

Gasification is an important technology to produce heat for thermal applications and power generation. Although, this is a century-old technology, effective use of gasifiers for power generation is yet not a widely accepted practice. The restricted use of gasifiers may be attributed to pollutants in producer gas such as tar and PM. However, as this report reveals, there are multiple methods suggested by various researchers to diminish pollutants. Of course, they also have their own pros and cons. Physical methods can reduce tar and PM but for effective cleaning, the product gas has to pass through multiple physical systems. Each incremental step in physical method increases pressure drop leading to reduce overall efficiency of system. Thermal cracking is not an economically viable option as a part of producer gas is utilized for cleansing of tar. Steam gasification has good potential to diminish tar but it is preferable only in case of excess availability of steam. Hence, use of catalyst seems to be the most prudent way for power generation from biomass from 3E (Environmental, Economic, and Efficiency) perspective. It has the potential to reduce tar as well as to increase conversion efficiency which may lead to increase in calorific value of producer gas.

References

1. Renewables (2014) Global status report, p 5
2. REN21 The First Decade (2004–2014) 10 years of renewable energy process. Renewable energy policy network for 21st century, p 9
3. Energy Statistics (2015) Central Statistics Office, Government of India, p 20
4. Govindjee WJ (1999) The photosynthetic process. In: Concepts in photo biology: photosynthesis and photo morphogenesis, Kluwer Academic Publishers, Boston, pp 11–51
5. Piriou B, Vaitilingom G, Veyssière B, Cuq B, Rouau X (2013) Potential direct use of solid biomass in internal combustion engines. Prog Energ Combust Sci 39:169–188
6. Deshmukh R, Jacobson A, Chamberlin C, Kammen D (2013) Thermal gasification or direct combustion, comparison of advanced cogeneration systems in the sugarcane industry. Biomass Bioenerg 55:163–174
7. Kumar A, Demirel Y, Jones DD, Hanna MA (2010) Optimization and economic evaluation of industrial gas production and combined heat and power generation from gasification of corn stover and distillers grains. Bioresour Technol 101:3696–3701
8. Kalina J (2011) Integrated biomass gasification combined cycle distributed generation plant with reciprocating gas engine and ORC. Appl Therm Eng 31:2829–2840
9. Leibbrandt NH, Aboyade AO, Knoetze JH, Görgens JF (2013) Process efficiency of biofuel production via gasification and Fischer–Tropsch synthesis. Fuel 109:484–492
10. Kim K, Kim Y, Yang C, Moon J, Kim B, Lee J (2013) Long-term operation of biomass-to-liquid systems coupled to gasification and Fischer–Tropsch processes for biofuel production. Bioresour Technol 127:391–399
11. Di Blasi C (1997) Influences of physical properties on biomass devolatilization characteristics. Fuel 76:957–964
12. Xu C, Donald J, Byambajav E, Ohtsuka Y (2010) Recent advances in catalysts for hot-gas removal of tar and NH3 from biomass gasification. Fuel 89:1784–1795
13. Šulc J, Štojdl J, Richter M, Popelka J, Svoboda K, Smetana J (2012) Biomass waste gasification—can be the two stage process suitable for tar reduction and power generation. Waste Manag 32:692–700
14. Shen Y, Yoshikawa K (2013) Recent progresses in catalytic tar elimination during biomass gasification or pyrolysis—a review. Renew Sustain Energ Rev 21:371–392
15. Paethanom A, Nakahara S, Kobayashi M, Prawisudha P, Yoshikawa K (2012) Performance of tar removal by absorption and adsorption for biomass gasification. Fuel Proc Technol 104:144–154
16. Chiang K-Y, Lu C-H, Lin M-H, Chien K-L (2013) Reducing tar yield in gasification of paper-reject sludge by using a hot-gas cleaning system. Energy 50:47–53
17. Milne TA, Evans RJ, Abatzaglou N (1998) Biomass gasifier "Tars": their nature, formation, and conversion. NREL Technical Report (NREL/TP-570-25357), National Energy Laboratory, Colorado, United States
18. Devi L, Ptasinski KJ, Janssen FJJG (2003) A review of the primary measures for tar elimination in biomass gasification process. Biomass Bioenerg 24:125–140
19. Machin EB et al (2015) Tar reduction in downdraft biomass gasifier using a primary method. Renew Energ 78:478–483
20. Hasler P (1997) Evaluation of gas cleaning technologies for small scale biomass gasifiers
21. Bridgwater AV (1995) The technical and economic feasibility of biomass gasification for power generation. Fuel 74:631–653
22. Woolcock PJ, Brown RC (2013) A review of cleaning technologies for biomass-derived syngas. Biomass Bioenerg 52:53–84
23. Bosmans A, Wasan S, Helsen L (2013) Waste to clean syngas: avoiding tar problems. In: 2nd international enhanced landfill mining symposium, Houthalen-Helchteren,14–16 Oct 2013

24. Chiranjeevaraoseela VC, Vykunta RM (2015) Techniques of tar removal from producer gas—a review. Int J Innov Res Sci Eng Technol, Department of Mechanical Engineering, GMRIT Rajam, India, Feb-2015
25. Hasler P, Nussbaumer Th (1999) Gas cleaning requirements for IC engine applications fixed bed biomass gasification. Biomass Bioenerg 16:385–395
26. Ma Z, Zhang Y, Zhang Q, Zhou J, Wang H, Tong P (2014). Design and experimental study of a two-stage pilot scale biomass fixed bed gasifier. College of Chemical Engineering, Nanjing Forestry University, china
27. Hindsgaul C (2000) Low temperature particle filtration of producer gas with low tar content. ET-ES 2000–2005
28. Neubauer Y (2011) Strategies for tar reduction in fuel-gases and synthesis-gases from biomass gasification. J Sustain Energ Environ Spec Issue 2011:67–71
29. Hofbauer H, Rauch R, Foscolo1 P, Matera D (2000) Hydrogen rich gas from biomass steam gasification. Institute of Chemical Engineering, Fuel and Environmental Technology Getreidemarkt 9/159, A-1060 Vienna
30. Han J, Kim H (2008) The reduction and control technology of tar during biomass gasification/pyrolysis: an overview. Renew Sustain Energ Rev 12(2):397–416
31. Li C, Suzuki K (2009) Tar property, analysis, reforming mechanism and model for biomass gasification—an overview. Renew Sustain Energ Rev 13(3):594–604
32. Bergman PCA, van Paasen SVB, Boerrigter H (2002) The novel "OLGA" technology for complete tar removal from biomass producer gas. Pyrolysis and gasification of biomass and waste, expert meeting, 1 Oct, Strasbourg, France
33. Laurence LC, Ashenafi D (2012) Syngas treatment unit for small scale gasification—application to IC engine gas quality requirement. J Appl Fluid Mech 5(1):95–103
34. Mandwe DS, Gadge SR, Dubey AK, Khambalkar VP (2006) Design and development of a 20 kW cleaning and cooling system for a wood-chip gasifier. J Energ South Afr 17(4):65
35. Nakamura S, Siriwata U, Yoshikawaa K, Kitanob S (2015) Development of tar removal technologies for biomass gasification using the by-products. Energ Procedia 75:208–213
36. Asadullah M (2014) Biomass gasification gas cleaning for downstream applications: a comparative critical review. Renew Sustain Energ Rev 40:118–132
37. Anisa S, Zainala ZA (2011) Tar reduction in biomass producer gas via mechanical, catalytic and thermal methods: a review. Renew Sustain Energ Rev 15:2355–2377
38. Surjosatyo A, Vidian F, Nugroho YS (2010) A review on gasifier modification for tar reduction in biomass gasification. J Mech (31):62–77
39. Ahmadi M, Sjosrtom K, Brage C, Liliedahl T (2015) In: Development of an online tar measuring method for quantitative analysis of biomass producer gas. SE-10044 Stockholm, Sweden
40. Akudo CO, Theegala CS (2014) Quantification of tars, particulates and higher heating values in gases produced from biomass gasifier. Tars Gasification BioResour 9(3):5627–5635
41. Ligang W, Shaoping X, Li Z, Changhou L, Hui Z, Shuqin L et al (2007) Steam gasification of biomass for hydrogen-rich gas in a free-fall reactor. Int J Hydrogen Energ 32:24–31
42. Franco C, Pinto F, Gulyurtlu I, Cabrita I, Edificio J (2003) The study of reactions influencing the biomass steam gasification process. Fuel 82:835–842
43. Saw WL, Pang S (2013) Co-gasification of blended lignite and wood pellets in a 100 kW dual fluidized bed steam gasifier: the influence of lignite ratio on producer gas composition and tar content. Fuel 112:117–124
44. Aitziber E, Gartzen L, Maider A, Javier B, Martin O et al (2014) Influence of operating conditions on the steam gasification of biomass in a conical spouted bed reactor. Chem Eng J 237:259–267
45. Hurley S, Xu CC, Preto F, Shao Y, Li H, Wang J, Tourigny G (2012) Catalytic gasification of woody biomass in an air-blown fluidized-bed reactor using Canadian limonite iron ore as the bed material. Fuel 91:170–176

46. Dong L, Asadullah M, Zhang S, Wang XS, Wu H, Li CZ (2013) An advanced biomass gasification technology with integrated catalytic hot gas cleaning part I: technology and initial experimental results in a lab-scale facility. Fuel 108:409–416
47. Zhang S, Asadullah M, Dong L, Tay HL, Li CZ (2013) An advanced biomass gasification technology with integrated catalytic hot gas cleaning. part II: tar reforming using char as a catalyst or as a catalyst support. Fuel 112:646–653
48. Li Yang, Yang H, Hu J, Wang X, Chen H (2014) Effect of catalysts on the reactivity and structure evolution of char in petroleum coke steam gasification. Fuel 117:1174–1180
49. Wang J, Xu S, Xiao B, Xu M, Yang L, Liu S, Hu Z, Guo D, Hu M, Ma C, Luo S (2013) Influence of catalyst and temperature on gasification performance of pig compost for hydrogen-rich gas production. Int J Hydrogen Energ 38:14200–14207
50. Liu H, Hu H, Luo G, Li A, Xu M, Yao H (2013) Enhancement of hydrogen production in steam gasification of sewage sludge by reusing the calcium in lime-conditioned sludge. Int J Hydrogen Energ 38:1332–1341
51. Chiang KY, Lu CH, Chien KL (2013) The aluminum silicate catalyst effect on efficiency of energy yield in gasification of paper-reject sludge. Int J Hydrogen Energ 38:15787–15793
52. Chiang KY, Lin YX, Lu CH, Chien KL, Lin MH, Wu CC, Ton SS, Chen JL (2013) Gasification of rice straw in an updraft gasifier using water purification sludge containing Fe/Mn as a catalyst. Int J Hydrogen Energ 38:12318–12324
53. Phuhiran C, Takarada T, Chaiklangmuang S (2014) Hydrogen-rich gas from catalytic steam gasification of eucalyptus using nickel-loaded Thai brown coal char catalyst. Int J Hydrogen Energ 39:3649–3656
54. Wang J, Xiao B, Liu S, Hu Z, He P, Guo D, Hu M, Qi F, Luo S (2013) Catalytic steam gasification of pig compost for hydrogen-rich gas production in a fixed bed reactor. Biores Technol 133:127–133
55. Sornkade P, Atong D, Sricharoenchaikul V (2013) Enhancement of cassava rhizome gasification using mono-metallic cobalt catalysts. Energ Procedia 34:273–281
56. Wang D (2013) Study of Ni/char catalyst for biomass gasification in an updraft gasifier: influence of catalyst granular size on catalytic performance. Bio Resour 8(3):3479–3489

Part IV
Biomass Management

Chapter 10
Partial Replacement of Aggregates Using Agricultural Waste in Concrete Manufacturing

Ranjeet Singh, Amit Arora, Gurmit Singh and Anand K. Tyagi

Abstract The environmental waste management practices are picking up across the world day by day. The present work deals with scientific disposal of the Agriculture waste material (rice straws) by utilizing them as coarse aggregate in conventional concrete mix. In order to reduce the alarmingly adverse effects of this waste material on the environment, it has become utmost necessary to get rid of it wisely so that the disposal may not affect the environment. The incorporation of this solid wastes in M15 grade concrete mix as partial replacements to coarse aggregate is found to yield an alternative lightweight eco-friendly concrete. The compressive strength of rice straws reinforced concrete sample with 10% replacement found out to be 10.2 N/mm^2 which is 68% as compared to conventional concrete i.e., M15 (15 N/mm^2). This waste utilization cum management practice makes the possibility of embedding the agriculture waste material into concrete, providing us a good sustainable, economical and lightweight alternative environment friendly construction material.

Keywords Compressive strength · Agriculture waste · Rice straws
Reinforced concrete

R. Singh (✉)
Department of Agriculture Engineering, Government Polytechnic Sirsa, Sirsa, Haryana, India
e-mail: ranjeetagri@hotmail.com

A. Arora
Department of Chemical Engineering, Shaheed Bhagat Singh State Technical Campus, Ferozepur, India

G. Singh
Punjab Pollution Control Board, Regional Office, Faridkot, India

A.K. Tyagi
Department of Applied Science & Humanities, Shaheed Bhagat Singh State Technical Campus, Ferozepur, India

S. Kumar et al. (eds.), *Conference Proceedings of the Second International Conference on Recent Advances in Bioenergy Research*, Springer Proceedings in Energy, https://doi.org/10.1007/978-981-10-6107-3_10

Highlights

1. It is a green concrete based on waste utilization
2. It can be used for lightweight concrete manufacturing
3. It gives detrimental effects in compressive properties but the reinforced concretes were found to have acceptable strengths

10.1 Introduction

Concrete is the most widely used building material in the world. Its composition contains aggregates, cement, and water [3]. The Chemical reaction between these components finally forms a hardened rocks like material, known as concrete. The hardening of the concrete depends on the properties of its components and composition. During the last century, the ordinary Portland cement, the conventional aggregates and mild steel, as reinforcement have been mainly used to attain the desired durability and utility of concrete. However, with the changed utilization patterns of these concrete components in light of the boom in real estate sector and continuously depleting aggregate sources, the research for the alternative aggregate materials has become a need of the hour. There is a strong possibility of utilizing the waste material like an agricultural waste as an alternatives to aggregates in concrete manufacturing, if not completely; at least partially as their disposal constituted to be an environmental challenge [2, 7].

The Environmental pollution associated with the aforesaid waste has lead us to think about the necessity to utilize or dispose-off these wastes in a manner so that their ill effects on the living being, flora and fauna may be avoided. After exploring a number of possibilities, we reached at a conclusion that these may be used as a partial replacement of aggregates in concrete control mix. So many studies being done to investigate their impact on the concrete manufacturing. The durability, compressive strength, and other characteristics along with the workability of this alternative material could be of greater interests of the researchers. This approach expected to be doubly beneficial; on the one hand, it will address the problem of increased cost of construction material and on the other it will be a safeguard to the environment, as it will be utilizing the solid waste material of hazardous nature [8]. Hence, there is a strong need to search for such ways by which these wastes can be used efficiently to reduce their harmful effects on the environment [1].

The prime motto of this study is to find out whether the waste materials that are harmful to the environment can be used as substitutes for normal aggregates in concrete and can thus be helpful in developing an alternative building material, probably with some improvement in properties [5]. The complete replacement of aggregates by these waste materials is not possible, therefore, researchers, throughout the world, are trying to find out different ways and means for changing the composition of concrete mix to make construction more environment friendly, sustainable and economic with possible improvement in some characteristics, if not all. A number of research groups have shown greater interest toward this goal of saving the environment through exploring the possibilities of utilizing the solid waste materials as a substitute to aggregates in concrete, thereby, initiating altogether a new solid waste management practice along with making the constructions more sustainable, economic, and environment friendly [6].

In the present study, we have attempted to use this agriculture waste (rice straws) in the conventional concrete mix, as a partial replacement of coarse aggregates. Four test specimens of size 0.1 m × 0.10 m × 0.1 m were prepared for testing. Out of these, one specimen was prepared by using conventional concrete M15 composition and three specimens were prepared by replacing the conventional coarse aggregates in it by weight of 10, 15, and 25%. All the test specimens were cured for 28 days under the controlled conditions [4]. After 28 days, all these specimens were tested on the compression testing machine to determine the compressive strength of the concrete. The characteristics of the (rice straws) concrete were then compared with the compressive strength of conventional M15 concrete sample.

10.2 Materials and Methods

The following materials were used as constituents of the concrete mix.

- Cement—Ordinary Portland Cement of 43 Grade was used in the study. The same was available under the commercial name of "Abuja cement".
- Sand—Coarse sand was used in this study.
- Aggregates—Aggregates used in the study obtained from the nearby quarry and are about 10 mm in size.
- Agriculture (Rice Straws)—the rice straws were brought from the local farm situated in Sirsa, District, Haryana. It is light in weight and light brown in color.

10.2.1 Preparation of Samples

In all four cubical concrete samples were fabricated, one conventional concrete mix specimen (Sample No-I) and three other specimens (Sample II, III, and IV) with replacement of aggregates by rice straws by weight of 10, 15, and 25%, respectively. All these test specimens were prepared, organized and tested as per the provisions of Indian standard IS-156-1959. The proportioning of various components of concrete mixture in different samples is as follows:

Sample No. I: It consists of a conventional concrete mix having cement, coarse aggregates (sand), coarse aggregates as its ingredients. Proportioning of various constituents by weight is shown in Table 10.1.

Sample No. II, III, and IV: These samples consist of a conventional concrete mix with replacement of coarse aggregates by Rice Straws. The samples were

Table 10.1 Proportioning of constituents of conventional concrete sample (by weight)

OPC cement	Coarse sand	Coarse aggregates	River water
0.15 kg	0.3 kg	0.6 kg	0.075 kg

Table 10.2 Proportioning of constituents of rice straws concrete sample (by weight)

Sample No.	Replacement (%)	Constituents				
		Cement (kg)	Fine aggregates (kg)	Coarse aggregates (kg)	Agriculture waste (rice straws)	Water (kg)
II	10	0.15	0.3	0.54	0.06	0.075
III	15	0.15	0.3	0.51	0.09	0.075
IV	25	0.15	0.3	0.45	0.15	0.075

Fig. 10.1 Image of agriculture rice straws based concrete

prepared by throughly bending the rice straws with the other starting materials: cement, fine and course aggregates, and water. Three specimen samples have been prepared for testing, by replacing the coarse aggregates by 10, 15, and 25% by weight. Proportioning of various constituents is shown in Table 10.2 and image of samples is as shown in Fig. 10.1.

All the four test specimens were casted using a cubical cavity mould of size ($100 \times 100 \times 100$) mm using the above-mentioned concrete mixes as per the provisions of IS-516-1959 and thereafter these were seasoned for 28 days under the stipulated conditions and tested on compressive testing machine as shown in Fig. 10.2.

10.2.2 *Testing of Cubical Specimens (Compressive Strength Test)*

All the specimens were tested on the compressive strength test machine by applying the load gradually without any jerk. The maximum load at which failure of the

Fig. 10.2 Compressive strength testing machine

specimen takes place is then recorded, and compressive strength of the cube specimen is measured by using the formula;

$$\text{Compressive strength}\left(\text{N/mm}^2\right) = W_f/Ap \tag{10.1}$$

where;

W_f Maximum applied load just before the load (kN)
Ap Plan area of cube mould (mm^2)

10.3 Result and Discussion

The conventional concrete mix cube (Sample I) exhibited the compressive strength of 15 N/mm^2, which is an obvious result of M15 grade mix design conventional concrete. The results of compressive testing of all the three samples II, III, and IV that were prepared with 10, 15 and 25% replacement of the conventional coarse aggregates with rice straws are summarized in Table 10.3 and Fig. 10.3.

Table 10.3 Compression properties of agriculture rice straws based concrete

Sample No.	Percentage replacement (by weight)	Compressive strength (at 28 days) N/mm^2
II	10	10.2
III	15	5.1
IV	25	3.4

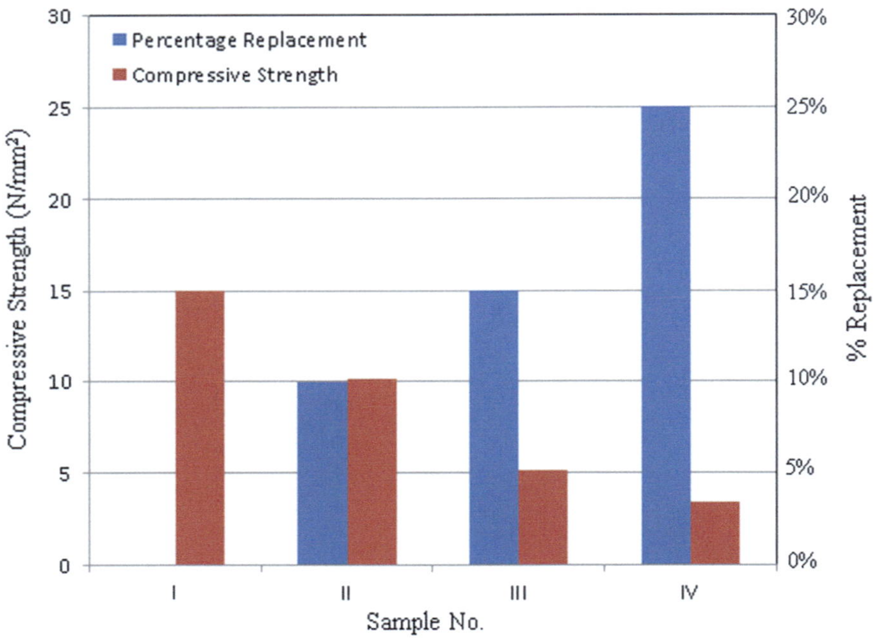

Fig. 10.3 Compressive strength of rice straws mix concrete (N/mm^2)

The compressive strength of the rice straw reinforced concrete sample with 10% replacement of aggregates is found out to be 10.2 N/mm^2 and at 15% replacement it was recorded to be 5.1 N/mm^2, which is about one third of the sample I. Likewise, as we move from sample III to IV, it reached to 3.4 N/mm^2, which is again boiled down appreciably by 1.7 N/mm^2.

It signifies that as we go from 10 to 25% replacement of course aggregates by rice straw in M15 concrete, the compressive strength of reinforced concrete samples decreases nearly monotonously with increase in percentage replacement of coarse aggregates as shown in Fig. 10.3. Thereby, it is evident that the reinforcement of rice straw in M15 concrete reduces the compressive strength. On the other hand, the fabricated reinforced concrete samples were found to of lightweight as evidenced from the density calculations. The reduction in density is because of the increased porosity due to embedded rice straw in the concrete matrix. Therefore, these materials are highly suitable for applications where strength is not the sole critarial for selecting the construction material. The reduced density and enhanced porosity observed in case of rice straw reinforced concretes make them a potential candidate material for flooring or pavement applications where water channels are required for draining out the rain water. This will obviously be having two fold advantage as far as the environment is concerned on the one hand we are using the hazardous solid waste and on the other we are utilizing the rain water also for maintaining the water table.

10.4 Conclusions

The present investigation was intended to find the effective ways to utilize the agriculture waste (rice straws) as aggregates of concrete manufacturing. Following are the findings of this study

1. The study revealed that the Agriculture waste (Rice Straw) can be incorporated as a coarse aggregate replacement in concrete. Though, it gives detrimental effects in compressive properties but the reinforced concretes were found to have acceptable strengths and potential development properties specially for a lightweight concrete manufacturing.
2. This eco-friendly and economic novel concrete strengthens the waste management practices on the one hand and opens up a new era of water absorbing concrete for hardening the rain water in applications like pavements and backyard flooring, on the other hand.

References

1. Adam B, Richard G, Robert B (2005) A tale of two market failure; technology and environmental policy. TCE 54:164–174
2. Bedi R, Chanda R (2013) Mechanical properties of polymer concrete. J Compos 17:42–61
3. Darji A, Patel I, Shah J (2014) Effects of green supplementary material on engineering properties of recycled aggregate concrete. JCEET 1(4):1–4
4. Ganesan K, Rajgopal K, Thangavel K (2007) Evaluation of bagasse ash as supplementary cementious material, ecological economics. Cement Concr Compos, JCCC 29:515–524
5. Kumar RS (2012) Experimental study of the properties of concrete made with alternate construction materials. IJMER 2(5):3006–3012
6. Poveda CA, Young R (2015) Potential benefits of developing and implementing environmental and sustainability rating system; making the case for the need for diversification. IJSBE 4(1):1–11
7. Sarath Chandra Kumar B et al (2004) Sustainable development using supplementary cementitious materials and recycled aggregates. IJMER 2:165–171
8. Tomas U, Jr Ganiron (2015) Recycling concrete debris from construction and demolition waste. IJAST 77:7–24

Chapter 11
Characterization of Wood Apple Shell Particles

O. Shakuntala, G. Raghavendra and S.K. Acharya

Abstract Characterization of bio waste fiber/particulates for the fabrication of bio composite is essential, because the performance of composite depends upon several factors, including chemical composition, physical properties, and environmental condition of fiber/particulates. In this research, a new bio waste lignocellulosic wood apple shell particles have been chosen and characterized their properties through various analyses such as proximate, Brunauer–Emmett–Teller (BET), energy dispersive spectroscopy (EDS), and X-ray diffraction (XRD), whether it is a suitable reinforcement for fabrication of polymer composite. From the analysis, it is concluded that the lignin percentage is more which helps to increase the carbon percentage in the shell particles. EDS result shows that wood apple shell contains hard particles which are suitable for the fabrication of polymer composite.

Keywords Wood apple shell · Proximate · BET · EDS

11.1 Introduction

Green composites which were prepared from agro-based bio fibers have rapidly expanded due to the availability and possess environmentally beneficial properties such as biodegradability. Depleting natural resources, emerging environmental awareness, and economic considerations are the major dynamic forces to exploit renewable resources such as agro-based fibers [1]. These lignocellulosic agricultural by-products that are available in considerable quantity and at low-cost are bamboo,

O. Shakuntala (✉) · G. Raghavendra
Department of Mechanical Engineering, NIT Warangal, Warangal, Telengana, India
e-mail: shaku30@gmail.com

S.K. Acharya
Department of Mechanical Engineering, NIT, Rourkela 769008, Odisha, India

S. Kumar et al. (eds.), *Conference Proceedings of the Second International Conference on Recent Advances in Bioenergy Research*, Springer Proceedings in Energy, https://doi.org/10.1007/978-981-10-6107-3_11

jute, kneaf, oil palm shell, coconut shell and sisal, etc., are subdivided based on their origins, coming from plants, animals, or minerals. Bio fibers have the composition, physical properties, and structure which are taking more attention in recent years in various applications including automotive, merchandise, structural, textile, and infrastructure. Over the last century, a number of researchers have been involved in exploring the utilization of bio fibers as load-bearing constituents in various composite materials. The use of these lignocellulosic fibers is being highly recommended using as reinforcing material in polymer composite because firstly, they are derived from plants and secondly, they are reusable, low density, flexibility during processing, desirable fiber aspect ratio, high specific stiffness, good insulator, biodegradability, less abrasive, good thermal properties and have a low-cost [2–3, 4]. On the other hand, the strength of lignocellulosic fiber mostly depends upon the cellulose, hemicelluloses, and lignin content. In case of lignocellulosic material, cellulose is a major structural component of plant cell walls and consisting of D-anhydroglucose ($C_6H_{11}O_5$) repeating units joined by 1,4-β-D-glycosidic linkages at C1 and C4 position [5]. Also, it provides better mechanical strength and chemical stability to plants. Hemicelluloses are a copolymer of different C5 and C6 sugar that also exist in the plant cell wall. Lignin is a polymer of aromatic compound product through a biosynthetic process and forms a protective layer for the plant walls [6].

In the present investigation, researcher has introduced a new lignocellulosic material, i.e., wood apple shell, and characterized their physical properties. Also, compared with other available literature for better understanding, which is most vital for fabricating a polymer composite.

11.2 Experimental Details

11.2.1 Material

Wood apple (Aeglemarmelos) belongs to family Rutaceae which is highly reputed medicinal tree commonly known as the 'Stone apple' or 'Bael' shown in Fig. 11.1. It is an indigenous fruit of India and found abundantly from sub-Himalayan forest, Bengal, central and south India. Among the different lignocellulosic fibers, wood apple shell appears to be promising material because of the high hardness and toughness.

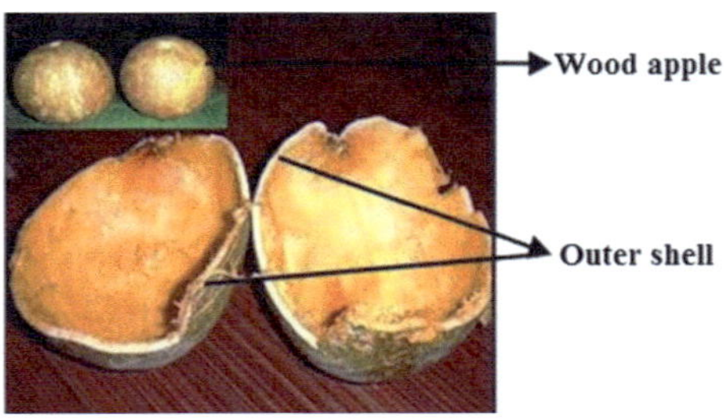

Fig. 11.1 Photomarcograph of wood apple shell

11.2.2 *Methods for Preparation of Shell Particles*

The wood apple shells used in the present investigation are initially washed with water to remove the impurities and dried in an oven at 110 °C for 24 h to remove the excess water content and moisture. The dried shells are converted to a fine powder using a ball milling process for 24 h and followed a sieve analysis to measure a particle size. The shell particles having density 1.068 g/cc chosen for the experiments are in the range of 212–1 μm as shown in Fig. 11.2.

11.2.3 *Characterization of Wood Apple Shell Particles*

11.2.3.1 Cellulose, Hemicellulose, and Lignin

The cellulose, hemicelluloses, and lignin of wood apple shell particles were determined by the method described by Abdul Khalil et al. [7]. The composition (α-cellulose, hemicelluloses, lignin, and ash contents) of the shell particulates was analyzed according to Technical Association of Pulp and Paper Industry (TAPPI) standards. The holocellulose content (α-cellulose + hemicellulose) of the wood apple shell particles was determined by treating the shell materials with a $NaClO_3$ and NaOH mixture solution. The difference between the values of holocellulose and α-cellulose gives the hemicellulose content of the shell material. The lignin content of the shell particle was found by treating them with a sulfuric acid solution based on TAPPI standard T222 om-83.

Table 11.1 compares the chemical composition of wood apple particulates with some other important lignocellulosic fibers. The cellulose content is greater than

Fig. 11.2 Photomacrograph of the sieve raw wood apple shell particles

Table 11.1 Chemical composition of different fibers

Species	α-Cellulose (wt%)	Hemicellulose (wt%)	Lignin (wt%)	Ash (wt%)	Ref.
Wood apple shell	39.54	26.06	30.86	0.9	
Coconut shell	43.0	13.3	45.0	2.2	[7]
Groundnut shell	35.7	18.7	30.2	5.9	[8]
Oil palm shell	65	–	19	2	[9]

groundnut shell. The hemicellulose content of wood apple shell is greater than other species. The lignin content of the wood apple shell is much greater than groundnut shell and oil palm shell. Combination of cellulose and hemicelluloses in wood apple is more when compared to other natural fibers.

11.2.3.2 Proximate Analysis

Initial characterization of the filler was done to give an insight into the expected behavior of filler material. The proximate analysis of wood apple shell particles has been done under the ASTM standards E-871, E-1755, E-872 for moisture, ash, and volatile matter, respectively. In addition to this, fixed carbon content was calculated by difference using the balance which is shown in Table 11.2.

$$\%FC = 100 - (\%ASH + \%VM + \%M), \quad (1)$$

where %FC, %ASH, %VM, and %M, respectively, indicates the weight percentages of fixed carbon, ash, volatile matter, and moisture of wood apple shell particles.

11.2.3.3 Ultimate Analysis

The ultimate analysis wood apple shell particulate was carried out in a CHNS Analyzer (CHN-932). The ultimate analysis of a sample determines the elemental compositions (carbon (C), hydrogen (H), nitrogen (N), and sulfur (S) contents) of the sample. It is based on the principle of the Dumas method which involves the complete and the instantaneous oxidation of the sample by flash combustion. From these results, the oxygen composition is determined by subtracting the sum of carbon, hydrogen, nitrogen, and sulfur compositions from 100. Table 11.3 shows the ultimate analysis of wood apple shell particles.

Table 11.2 Proximate analysis of wood apple shell particulates

Material	Proximate analysis (wt% by mass basis)			
	FC	Moisture	VM	ASH
Wood apple shell	19.21	6.6	73.34	0.95

Table 11.3 Ultimate analysis of wood apple shell particulates

Wood apple shell	Elemental analysis (wt% by mass basis)				
	C	H	N	S	O
Raw particulates	66.387	6.859	0.633	0	25.278

11.2.3.4 Elemental Analysis

The elemental composition of wood apple shell particles is determined by energy dispersive spectroscopy (EDS). The EDS was obtained in a 'spot mode' in which beam is focused on a single area selected within the field of view.

11.2.3.5 BET Analysis

The BET surface area was determined from the adsorption isotherms using the Brunauer–Emmett–Teller (BET).The Dubinin–Radushkevich (DR) equation was used to calculate the surface area of the wood apple shell particles which is shown in Fig. 11.4.

11.2.3.6 XRD Analysis

XRD is the most common technique used to characterize the crystalline content in a polymer structure. X-ray diffraction studies have been widely used in particle research to characterize their critical features such as crystallite size, chemical composition, and strain. The peak positions are indicative of the crystal structure and symmetry of the contributing phase. The ball-milled powder was characterized with an X-ray diffractometer (PAN alytical, PW3050/60 XRD; Cu Kaanode; $k = 0.154$ nm). X-pert software was used to investigate the structural changes and phase transformations of the micro-powders. The samples were gently consolidated in an aluminum holder and scanned at 45 kV and 40 Ma. The samples were scanned in the range from 10° to 90° of 2θ with a scan rate of 20/min and a step size of 0.06.

11.2.3.7 Scanning Electron Microscopy (SEM)

The fracture surfaces of the composite specimens are examined directly by scanning electron microscope JEOL JSM-6480LV. The samples are stacked on stubs with silver paste.

11.3 Results and Discussion

Figure 11.3 exhibits the inspection spectra of wood apple shell particulate surface elements acquired for wood apple shell. The particulates surface exhibit spectra containing mainly carbon, oxygen, silicon, aluminum, small amount of zirconium

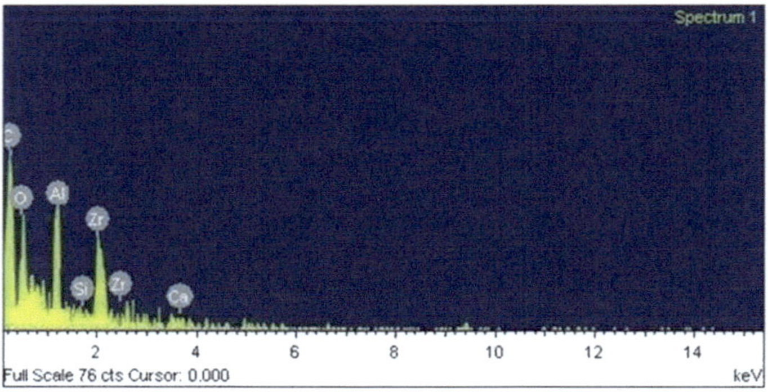

Fig. 11.3 EDS of wood apple shell particles

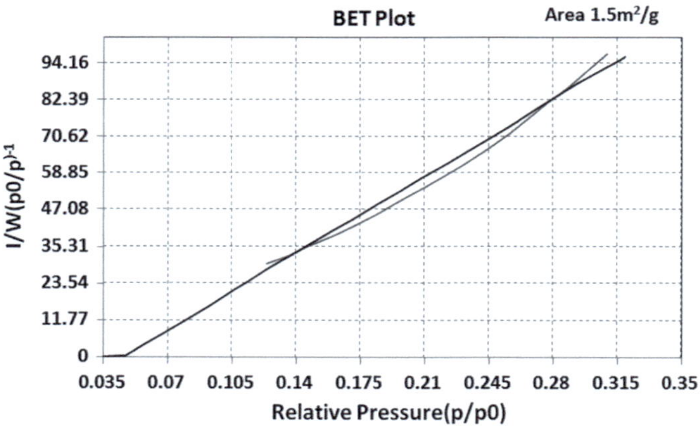

Fig. 11.4 BET analysis of wood apple shell particles

and calcium. Due to the presence of hard elements like SiO_2 and Al_2O_3, the raw wood apple shell powder can be used as particulate reinforcement in various polymer matrixes.

From the BET analysis, the surface area obtained for the wood apple shell particles is 1.5 m^2/g which is shown in Fig. 11.4.

Figure 11.5 shows the XRD pattern of the wood apple shell particles in which the major diffraction peaks are 23.56°, 26.47°, 34.68°, 38.40°, 41.45°, and 53.48°. Phases of these peaks as SiO_2, Mg_2, C, MgO, Al_2O_3 revealed that this particle has some of the composition of hemicelluloses, cellulose, and lignin that has been confirmed by the literature [10].

The scanning electron macrograph of wood apple shell particles is shown in Fig. 11.6. The wood apple shell particulates come under the range of 1–212 μm.

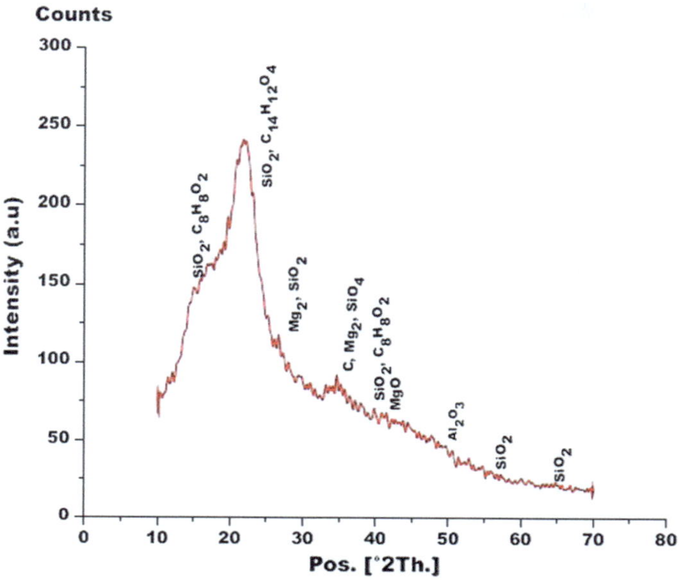

Fig. 11.5 XRD pattern of the wood apple shell particles

Fig. 11.6 SEM of wood apple shell particles

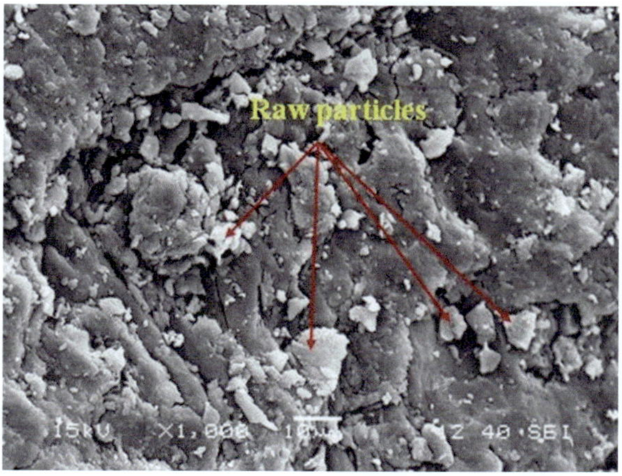

11.4 Conclusions

Characterization of the new wood apple shell gives brief description regarding the composition and physical properties of the shell particles. Cellulose, hemicellulose, and lignin content are better as compared to other fiber. If lignin content is more in the particles, then the carbon percentage is also maximized, which is observed through proximate and ultimate analysis. Due to the presence of hard elements like

SiO_2 and Al_2O_3, the wood apple shell particles can be used as particulate reinforcement, which increases the strength and hardness property of various polymer matrixes. Hence, this characterization is most probably helpful for the fabrication of biopolymer composite.

References

1. Reddy N, Yang Y (2005) Biofibers from agricultural byproducts for industrial applications. Trends Biotechnol 23(1):22–27
2. Ojha S, Acharya SK, Raghavendra G (2014) A novel approach to utilize waste carbon as reinforcement in thermoset composite. Proc Inst Mech Eng Part E J Process Mech Eng p 0954408914547118
3. Mwaikambo LY, Bisanda ETN (1999) Performance of cotton-kapok fabric-polyester composites. Polym Test 18(3):181–198
4. Joseph K, Varghese S, Kalaprasad G et al (1996) Influence of interfacial adhesion on the mechanical properties and fracture behaviour of short sisal fibre reinforced polymer composites. EurPolym J 32(10):1243–1250
5. John Maya Jacob, Thomas Sabu (2008) Biofibres and biocomposites. Carbohyd Polym 71(3):343–364
6. Harmsen P, Huijgen W, Bermudez L, Bakker R (2010) Literature review of physical and chemical pretreatment processes for lignocellulosic. Biomass 1–49
7. Abdul Khalil HPS, Firoozian P, Bakare IO, Akil HM, Noor AM (2010) Exploring biomass based carbon black as filler in epoxy composites: flexural and thermal properties. Mater Des 31:3419–3425
8. Raju GU, Kumarappa S (2011) Mechanical and physical characterization of agricultural waste reinforced polymer composite. J Reinf Plast Compos 30:1029
9. Jacob M, Thomas S, Varughese KT (2004) Mechanical properties of sisal/oil palm hybrid fiber reinforced natural rubber composites. Compos Sci Technol 64:955–965
10. Hattotuwa G, Premalal B, Ismail H, Baharin A (2002) Polym Test 21(7):833–839

Chapter 12
Governance of Sustainable Agriculture Schemes in India with Special Reference to Bio-fertilizer Project and Analysis

Navreet and Ravneet Kaur

Abstract Chemical fertilizers are one of the significant elements which boost the green revolution in India in achieving the goals of self-sufficiency in food production. This positive outcome had also brought negative side effects such as salinity, decreases in soil fertility, insect resistance to pesticides etc., which had further affected environment to large extent and had led to land degradation, water contamination, groundwater depletion and climate change. These problems in agricultural land and water resources could only be solved to some extent with usage of bio-fertilizers, bio-pesticides and adoption of organic farming techniques which are eco-friendly. The Government of India had implemented policies and programmes for improving the production, distribution, and promotion of bio-fertilizers. An attempt has been made to analyse the governance mechanism in improvising the market of bio-fertilizer in India while understanding the roles, capacities, and relationships among the diverse actors involved in governance process. It has been found that usage of bio-fertilizers is limited as compared to involvement of population in agriculture sector due to various governance issues in agriculture sector like low institutional capacity, lack of economic stability, poor accountability and transparency, no responsiveness, lack of participation and access to information.

Keyword Bio-fertilizer · Good governance · Sustainable agriculture
Soil salinity · Microbes

12.1 Introduction

Agriculture is a crucial sector for socio-economic development of India as the livelihood of majority of the country's population depends on agriculture. India being an agricultural country provides employment to 65% of the labour force,

Navreet (✉) · R. Kaur
Public Administration Department, Panjab University, Chandigarh, India
e-mail: navreet9@gmail.com

S. Kumar et al. (eds.), *Conference Proceedings of the Second International Conference on Recent Advances in Bioenergy Research*, Springer Proceedings in Energy, https://doi.org/10.1007/978-981-10-6107-3_12

accounts for about 14% of Gross Domestic Product and 43% of India's geographical area are used for agriculture activities [1]. Table 12.1 depicts the performance of the agricultural sector in India which clearly shows that GDP gained from agricultural sector in India is continuously decreasing while the population dependent on agriculture sector in India is continuously increasing from 1980 to 2015.

Indian agriculture sector has undergone dynamic change since the "Green Revolution", which provided self-sufficiency and ushered in an era of rural prosperity. The production of food grains had increased from 50.82 million tonnes in 1950–51 to 211.2 million tonnes with use of chemical fertilizers, high-yielding varieties, plant protection, chemicals and increase in irrigation facilities. The consumption of pesticides had increased from approximately 160 metric tonnes in 1949–50 to 64,000 metric tonnes (2002–03). There was a 250-fold increase in fertilizer consumption, from 69,000 tonnes during 1950–51 to 17.4 million tonnes of Nitrogen Phosphorous Potassium (NPK) nutrients during 2001–02, from 0.5 to 91 kg/ha. The excessive use of fertilizers and pesticides had negatively affected the nutrient composition in the soil and had created imbalance in it. Furthermore, nitrate concentration in groundwater and accumulation of heavy metals like arsenic, lead and cadmium had also been identified [3].

The excessive and inappropriate use of chemical pesticides had decreased the soil fertility to large extent and insect resistance to pesticide had further affected the environment. These had resulted in groundwater depletion, decreased soil fertility, land degradation, water contamination, groundwater depletion and climate change [4].

In order to meet the goals of sustainable development, sustainable agri-process is an essential condition for "inclusive growth". The process of agricultural development includes adoption of modern technology based upon usage of bio-fertilizer, bio-pesticides, results in positive effects for the sector and reduces the environment risks [5].

It has been found through research studies that combined usage of bio-fertilizers and chemical fertilizers helped to increase yield. Likewise if Phosphorous (P) chemical fertilizer is applied with other bio-fertilizers like Phosphate Solubilizing Bacteria (PSB), *Bacillus megatherium* var. phosphaticum, it increased the PSB population in the rhizosphere and P availability in the soil. It also enhanced sugarcane growth, its yield and quality. It was also stated that with combined treatment of multifunctional bio-fertilizer plus 50% chemical fertilizer on lettuce yield, there was a 25% increase of lettuce yield [6].

Table 12.1 Status of agriculture sector in India

Year	Agriculture share in GDP (%)	Population dependent on agriculture (%)
1980	39	70
1990	31	65
2000	25	59
2010	16	58
2015	13.7	65

Source Arjun [2]

12.2 Sustainable Development of Agriculture Through Usage of Bio-fertilizers

Sustainable development of agriculture sector through bio-fertilizer is the only solution which means to inculcate efforts to use the precious resource i.e. bio-fertilizers in Indian farms to preserve the environment and to meet the needs and expectations of both present and future generations from agriculture sector. "Sustainable agriculture can be defined as environmentally friendly methods of farming that allow the production of crops and livestock without damage to human or natural systems" [7].

Bio-fertilizers are products containing living cells of different types of micro-organisms, which have an ability to convert nutritionally important elements through biological processes [8]. Bio-fertilizers like Rhizobium, Azotobacter, Azolla are microbial inoculants, which fix atmospheric nitrogen either symbiotically with free living as well as phosphate solubilizing micro-organisms.

The microbial fertilizers have several advantages over conventional chemicals for agricultural purposes: (i) microbial products are considered safer than many of the chemicals now in use; (ii) neither toxic substances nor microbes themselves will be accumulated in the food chain; (iii) self-replication of microbes circumvents the need for repeated application; (iv) target organisms seldom develop resistance as is the case when chemical agents are used to eliminate the pests harmful to plant growth and (v) properly developed bio-control agents are not considered harmful to ecological processes or the environment [9].

Many countries in world had adopted the use of bio-fertilizers for agriculture. Europe and rest of the World countries especially Latin America are the two top consumers of bio-fertilizers in the world [10]. Europe and Latin America are the top consumers due to availability of stringent regulations imposed on chemical fertilizers, followed by Asia-Pacific which controlled 34% of the market as by 2011 [11]. Growth of bio-fertilizers is increasing in emerging countries such as China and India [10]. In 2011–12, the annual production of bio-fertilizer is recorded at about 40,324 tonnes in India. Maharashtra State tops in bio-fertilizers, producing 8743.69 tonnes followed by Uttar Pradesh (8695.08 tonnes) [10].

12.3 Overview of Bio-fertilizer Scheme in India

The Government of India implemented a central scheme called "National Project on Development and use of bio-fertilizers" (NPDB) for improving the production, distribution and promotion of bio-fertilizers which has resulted in growth of bio-fertilizers nearly 170 times in last 50 years after independence [10]. The scheme also covered training courses for extension workers and field demonstrations and providing quality control services [12]. Agro Industries Corporation, State Agriculture Departments, National Bio-fertilizers Development Centres, State

Agricultural universities and private bio-fertilizer companies are the vital players in bio-fertilizer market.

As shown in Table 12.2, the region-wise analysis depicts that distribution of bio-fertilizers in western region is highest followed by south and it is very less in northern region. On the other hand use of chemical fertilizer is highest in northern region.

A National Bio-fertilizer Development Centre was established at Ghaziabad as a subordinate office of the Department of Agriculture and Cooperation with six regional centres. Later on this scheme was merged in "The National Project on Organic Farming (NPOF)" which was launched during 10th Five Year Plan as pilot project with an outlay of Rs. 57.04 crore, it also continued in the 11th plan with an outlay of Rs. 101.00 crore [10]. These financial aids have resulted in increasing the production and distribution of bio-fertilizers in whole India. Currently, there are total 126 bio-fertilizer units engaged in bio-fertilizer production, and the government has extended financial assistance to 73 bio-fertilizer units for commercial production [3]. There are total 64 bio-fertilizers production units which have total installed production capacity of 900 tonnes and over 6000 tonnes of the total Bio-fertilizers production [13].

Table 12.3 shows the production of different bio-fertilizers in India. Among all the bio-fertilizer in India during 2006–07 production of PSB was observed maximum, 2920.42 tonnes. Azolla is relatively new in use, and its production was only 13.40 tonnes in 2006–07.

Table 12.4 shows that the production of bio-fertilizers has increased from 1992–93 to 2009–10. The total dispatches of bio-fertilizers had also increased from 1992–93 to 2009–10. The growth rate of both production and consumption of bio-fertilizer in India shows the positive trend but still there remain a huge gap between the installed capacity and production of bio fertilizer in India.

Table 12.2 Region-wise distribution of chemical and bio-fertilizers

Region	Chemical fertilizers (tonnes)	Bio-fertilizers (tonnes)
East	3417 (17%)	385.96 (3%)
West	5077 (25%)	5271.48 (46%)
North	6420 (32%)	544.44 (5%)
South	5425 (27%)	5155.74 (45%)

Source FAI quoted in Majumdar [6]

Table 12.3 Year-wise production of bio-fertilizer (tonnes) at all India level

Bio-fertilizer	2001–02	2003–04	2005–06	2006–07
Rizobium	1841.54	1822.94	1548.53	1579.66
Azotobacter	1324.74	1386.75	1874.93	2037.75
Azospirillum	1275.50	1159.26	1228.75	1283.70
Phosphate solubilizing bacteria	4502.22	4005.18	6075.52	6920.42
Azolla			25	13.40

Source FAI quoted in Majumdar [6]

Table 12.4 Installed capacity, production and despatches of bio-fertilizer in India

Year	Installed production capacity	Total production (tonnes)	Total dispatches (tonnes)	Unutilized capacity percentage
1992–93	5400.5	2005	1600.01	62.87
1993–94	6125.5	3084	2914	49.65
1994–95	8114.5	5800.5	4988.9	28.52
1995–96	10680.4	6692.3	6288.3	37.34
1996–97	12,647	7406.6	6681.4	41.44
1998–99	16,446	5972.1	5065.5	63.69
2001–02	15439.0	9019.2	8429.3	41.59
2002–03	18679.5	7181.7	7029.9	61.55
2003–04	18,632	8701.4	8375	53.30
2004–05	NA	10,479	10,428	–
2005–06	NA	11,752	11,358	–
2006–07	43,495	15,871	15,745	63.51
2007–08	67,162	20,111	20,100	70.06
2008–09	68,804	25,065	25,000	63.57
2009–10	68,078	20040.3	20,000	70.56

Source FAI quoted in Majumdar [6]

12.4 Governance of Bio-fertilizer Programmes

Administrative governance of schemes by government plays a significant role in implementation of that scheme to make it substantially reachable to beneficiaries. It has also been concluded through research studies that by 2020, to achieve the targeted production of 321 million tonnes of food grain, the requirement of nutrient will be 28.8 million tonnes, while their availability will be only 21.6 million tonnes being a deficit of about 7.2 million tonnes [14].

The governance system of food and agriculture in India suffers many issues which has affected the sustainable agriculture. This demands the development of new concept in agriculture sector of India i.e. Agriculture Governance. "It means augmentation of growth and development of a country's agriculture sector and managing the consequences of this process through the effective functioning of its institutions, the application of technology and scientific innovations, and the implementation of policies, adherence to acts and regulations, and active participation of all involved stakeholders" [15]. The studies have been conducted and it has been found that the application of microbial fertilizers on farms in practice has not achieved desirable effects. Key among the common characteristics which affects the governance process of bio-fertilizer projects includes:

- Inadequate or incomplete policies and guidelines for regulation of bio-fertilizers and bio-pesticides.
- Multiple and often overlapping regulatory mandates by government agencies.

- Limited capacity including staff, skills and laboratory for product monitoring.
- Inadequate enforcement of quality control for bio-fertilizers and bio-pesticides.
- Lack of bio-fertilizer and bio-pesticide specific regulations, standards and guidelines.
- Weak institutional arrangements with limited collaboration between relevant agencies [11].

12.5 Issues and Challenges in Governance of Bio-fertilizers at Country, State and Local Level

The analysis of various aspects of Bio-fertilizer scheme shows that annual requirement and production of different bio-fertilizers have shown tremendous gap. The region-wise analysis shows that northern region used least percentage of bio-fertilizers. There are various issues and challenges related with governance mechanism of bio-fertilizers. These are as follows:

- **Responsiveness**: The government machinery needs to be responsive towards maintaining a balance between various stakeholders involved in Bio-fertilizer scheme for the actual success of bio-fertilizer technology. Production and distribution are significant to sustain the demand and supply of bio-fertilizer, and research is significant to develop more efficient, temperature tolerant and hardy strains. Other elements under responsive behaviour like development of suitable carriers, better packaging and longer shelf life are also important for commercial acceptance of these living inputs by beneficiaries [16]. India has a vast agricultural extension system (Government Agriculture Department like State Agricultural Universities, Indian Council of Agricultural Research, input industries, Non-Governmental Organizations) which needs to be fully utilized to educate the farmers regarding bio-fertilizer. However, the reach and coordination of extension agencies to farmers are very weak, and the state extension machinery in reality almost does not exist. It leads to wide gap between potential and actual yields of farms while using bio-fertilizer [17].

It is government responsibility to make farmer's knowledgeable regarding the right product, dosage, time and method of application in accordance to soil and crop need, to reduce inefficient use of fertilizers. It is important to rejuvenate and reorient extension systems with focus on the poor farmers and low fertilizer consumption areas and to increase farm profitability (Planning Commission Report). The government needs to take a responsive step in encouraging and funding research by either universities or private firms to make up for shortfall in private commercial research initiative.

- **Accountability and Transparency**: Various institutions are involved in Bio-fertilizer programme like Central government (Ministry of Agriculture,

Department of biotechnology), Research and Development organizations (Indian Council of Agricultural Research, Council of Scientific and Industrial Research, State Agriculture universities) State government, several NGO's and large numbers of private sector and co-operative sector organizations. Sangar stated that "all these stakeholders faced suspicion about their accountability towards beneficiaries of bio-fertilizers due to insufficient extension efforts of government related to bio-fertilizers production and distribution mechanism and poor quality of the products due to poor handling, insufficient infrastructure for distribution and storage, limited shelf life, and climate specificity" [18]. There is need of evaluation and monitoring of working and performance of organizations involved in bio-fertilizer project by government to bring transparency and accountability. There can be equal provision of subsidy to farmers, distribution of annual booklets regarding production, distribution and research of bio-fertilizers to all stakeholders. The government needs to assure about the quality of bio-fertilizers by doing more rigorous research.

- **Participation of Stakeholders**: Participation of all stakeholders in implementation of any scheme is necessary requirement in order to have good results. The involvement of farmers is also necessary by getting themselves informed and knowledgeable of this new technique and methods to efficiently use bio-fertilizers in their farms [16]. These micro-organisms are extremely perishable and sensitive to local conditions and their proper handling by farmers also shambles its quality [12]. Bio-fertilizers can provide various opportunities like local employment through decentralized rural infrastructure, more improvised skills and capacities to address bio-technology, research and production capacities of soils [18]. It is also important for farmers to keep themselves update regarding their usage, quality, handling by attending seminars and conferences organized by Government, NGOs and various institutions at local and national level [16]. Participation in the form of exchange and interactions of individual farmers or farmer groups will support the usage of bio-fertilizer and leads to interactive learning and participatory technology development and transfer.
- **Dissemination and Promotion of Bio-fertilizers**: Bio-fertilizer technology can be promoted and disseminated among Indian farmers through demonstration and training modules. The other method is distribution of pamphlets, and other literature and organizing training modules for beneficiaries during farmer's meeting will enhance their knowledge and trust. Promotion through Kisan Mela can also be done. The promotion of bio-fertilizer by various social and governmental institutes will improve the participation level of beneficiaries.
- **Consensus Orientation and Strategic Vision**: These good governance principles indicate the essentiality of planning strategies, policies and procedure regarding promotion and spread of bio-fertilizer through consensus among different groups to make bio-fertilizers a vital part in Indian farms. In India mainly farming is largely practiced by number of small farmers who cannot afford high-priced chemical fertilizers. Small farmers are dependent on government subsidies and suffer from both soil quality deterioration and declining

yield response. Bio-fertilizers can only play a major role with the consent of farmers, companies and government to take Indian farms out of total dependence on harmful chemical fertilizers and contribute to long-term sustainability by strengthening of local production systems and improvement of natural resources [18]. The strategic vision of companies involved in producing, importing and marketing fertilizer and is basic requirement for Indian government and should also accurately forecast demand of bio-fertilizers by farmers. This will ensure timely supplies of the right bio-fertilizer types and maintain required storage, transport, staffing, credit, financial and foreign exchange arrangements. The dealership of bio-fertilizer should be upgraded to agro-input sale cum service centres by equipping them with diagnostic facilities, bringing all agro-inputs under one roof and providing linkages with banks for credit [17].

12.6 Conclusion

The green revolution had brought significant changes in agriculture sector in India. The phenomenal increase in production yields had also brought the environment challenges in form of land degradation, soil salinity, water contamination and depletion of ground water. There is urgent need to use the agricultural technology in order to move from production oriented to profit oriented sustainable farming. The conditions for development of sustainable agriculture are becoming more and more favourable with introduction of innovative technologies like Bio-fertilizer. National Project on Development of Bio-fertilizer scheme aims for giving grant up to Rs. 20 lakh per unit of 150 tonnes per year to set up bio-fertilizer producing units. Since inception of NPDB, bio-fertilizer production capacity of 10,525 tonnes has been envisaged by setting up 83 bio-fertilizer production units. Another 39 units have also been set up by different organizations and private entrepreneurs. The estimated total fertilizer production capacity in India is 18,500 tonnes per year, whereas estimated bio-fertilizer production is about 10,000 tonnes per year in the country.

The statistics show that the Bio-fertilizer scheme is unable to give fruitful results due to its ineffective approach towards its stakeholders especially beneficiaries. This requires a change in administrative system of agriculture sector. Fusion of Good governance principles in Bio-fertilizer scheme results in establishing effective mechanism of government institutions, improvise process in catering interest of farmers related to usage of bio-fertilizers. All the actors involved in governance process need to focus on incorporating the principles of good governance in promotion of bio-fertilizers to bring an efficient and sustainable output and outcome from agriculture sector. A clear administrative process will attract and allow businessmen to operate in a secure environment and can also attract new investors in the bio-fertilizer industry. This will deliver good yield in farms and boost the market of bio-fertilizers in India.

References

1. Sengupta A, Sonwani D (2012) Sustainable development in India with reference to agricultural sector. Int J Emerg Res Manag Technol. http://www.ermt.net/docs/papers/Volume_1/Issue_2_Decemebr2012/V1N1-0102.pdf. Accessed 1 Jan 2016
2. Arjun KM (2013) Indian agriculture status, importance and role in Indian economy. Int J Agric Food Sci Technol 4(4):343–346. http://www.ripublication.com/ijafst_spl/ijafstv4n4spl_11.pdf. Accessed on 29Aug 2016
3. Business Potential for Agricultural Biotechnology (2007) Report of the APO multi-country study mission on the business potential for agricultural biotechnology products. Asian Productivity Organization. http://www.apo-tokyo.org/publications/wp-content/uploads/sites/5/agr-19-bp_abp.pdf#page=99. Accessed 12 Jan 2016
4. Alam G (2000) A study of biopesticides and biofertilisers in Haryana, India. Gatekeeper series no. 93, International Institute for Environment and Development. http://pubs.iied.org/pdfs/6348IIED.pdf. Accessed 14 Jan 2016
5. Meghalaya Basin Development Authority, Sustainable Development of Agriculture: Integrated Basin Development and Livelihoods Programme, http://mbdaonline.com/PDFs/Sustainable_Agriculture.pdf. Accessed on 11 Jan 2016
6. Majumdar K (2015) Bio-fertilizer use in Indian agriculture. Indian J Res 4(6):377–381. http://worldwidejournals.com/paripex/file.php?val=June_2015_1434434436__125.pdf. Accessed 29 Aug 2016
7. Mishra P, Dash D (2014) Rejuvenation of biofertilizer for sustainable agriculture and economic development. J Sustain Develop 11:41–61. http://www.consiliencejournal.org/index.php/consilience/article/viewFile/350/176. Accessed 9 Jan 2016
8. Wu SC, Cao ZH, Li ZG, Cheung KC, Wong MH (2005) Effects of bio-fertilizer containing N-fixer, P and K solubilizers and AM fungi on maize growth: a greenhouse trial. Geoderma 125:155–166. http://ac.els-cdn.com/S0016706104001922/1-s2.0-S0016706104001922-main.pdf?_tid=8dbb0542-5aa1-11e5-b376-00000aab0f02&acdnat=1442208741_49748a56e831467b173a791a833f9663. Accessed 15 Jan 2016
9. Sharma J, Gurung T, Nandy K, Mitra AK (2014) Efficiency of different nitrogen fixing bacteria with respect to growth and development of legumes. Int J Curr Microbiol Appl Sci 3:799–809. http://www.ijcmas.com/vol-3-10/Juhi%20Sharma,%20et%20al.pdf. Accessed 14 Jan 2016
10. Kholkute R (2014) Bio-fertilizers: opportunities and challenges, Arab Fertilizer 65. http://www.ifaj.org/fileadmin/filedb/a/2014/20141128_RK_Biofertilizers_Opportunities_and_Challenges_.pdf. Accessed 12 Jan 2016
11. Simiyu NSW, Tarus D, Watiti J, Nangayo F (2013) Effective regulation of bio-fertilizers and bio-pesticides: a potential avenue to increase agricultural productivity, COMPRO II, policy series. http://www.aatf-africa.org/files/COMPRO-II_Policy-Brief%201_2013.pdf. Accessed 9 Jan 2016
12. Ghosh N (2004) Promoting bio-fertilizers in Indian agriculture. http://www.ipni.net/ipniweb/portal.nsf/0/94cfd5a0ed0843028525781c0065437e/$FILE/12%20South%20Asia.Ghosh.Promoting%20Bio-fertilizers%20in%20India%20Agri.pdf. Accessed 9 Sept 2015
13. Mazid M, Khan TA (2014) Future of bio-fertilizers in Indian agriculture: an overview. Int J Agric Food Res 3:10–23. https://www.sciencetarget.com/Journal/index.php/IJAFR/article/download/132/123. Accessed 19 Mar 2015
14. Ghumare V, Rana M, Gavkare O, Khachi B (2014) Bio-fertilizers-increasing soil fertility and crop productivity. J Ind Pollut Control 30:196–201. http://www.icontrolpollution.com/articles/biofertilizers-increasing-soil-fertility-and-crop-productivity-196-201.pdf. Accessed 19 Jan 2015
15. Dasgupta S, Roy I (2011) Good agricultural governance. Food and Agriculture Organization of The United Nations Regional Office for Asia and the Pacific. http://www.fao.org/3/a-ba0113e.pdf. Accessed 15 Sept 2015

16. Singh K (2015) Stimulating bio-fertilizers in Indian farming sectors. Indian J Appl Clin Sociol 10:1–18. http://www.ijacs.in/files/10%20(2)%20-%201.pdf. Accessed 9 Jan 2016
17. Planning Commission Report. http://planningcommission.nic.in/aboutus/committee/wrkgrp11/wg11_fertliser.pdf. Accessed 11 Feb 2016
18. Sangar S (2011) Enabling poverty relevant bio-fertilizer innovation systems—lessons from India. https://idl-bnc.idrc.ca/dspace/handle/10625/49219. Accessed 11 Feb 2016

Part V
Waste to Management

Chapter 13
Sequencing Batch Reactor Technique for Municipal Sewage Treatment with Carbon Credits

R.R. Marlar, Vigneshwaran Aiyappan, S.S. Rao and S. Bajpai

Abstract Sequencing batch reactor is a developing technology in wastewater treatment for municipal sewage and industrial effluent. Aerobic process treatment generates significant amount of greenhouse gases mainly carbon dioxide (CO_2) and nitrous oxide (N_2O). Reduction of GHG emissions is a significant role in wastewater treatment. The greenhouse gas (GHG) emissions from wastewater treatment is a significant consideration for the treatment process evaluation. In the present work, the direct and indirect emissions from a single-stage sequencing batch (SBR) deammonification plant for nitrogen elimination from sludge liquor are calculated based on measurements and construction documentation. The results show that N_2O is responsible for most CO_2 emissions, followed by energy consumption and construction and machinery, but significant amounts can be saved in mainstream treatment. The present work reveals that there is a need for the development of less GHG emitting process configurations. In this paper, an attempt has been made to evaluate the GHGs for municipal wastewater (mainly CO_2 and N_2O) with SBR technology for carbon credits approach. The analysed results are compared in a detailed manner. Neutral and negative carbon footprint has been widely discussed and investigated.

Keywords Sequencing batch reactor · Carbon credits
Municipal sewage treatment · Carbon footprint

13.1 Introduction

Municipal wastewater treatment processes are the emissions of the GHGs (like CH_4, N_2O and CO_2). GHGs are the causes for the climatic changes also [1]. The problems raised in conventional wastewater treatments are GHG emissions.

R.R. Marlar (✉) · V. Aiyappan · S.S. Rao · S. Bajpai
Department of Chemical Engineering, Dr. B R Ambedkar National Institute of Technology, Jalandhar, India
e-mail: robinstillrules@gmail.com

S. Kumar et al. (eds.), *Conference Proceedings of the Second International Conference on Recent Advances in Bioenergy Research*, Springer Proceedings in Energy, https://doi.org/10.1007/978-981-10-6107-3_13

Conventional treatment needs lots of aeration and electrical energy that effects the GHG emissions also. This aerobic treatment also causes large quantities of biosolids which need to be further treated [2]. Among these GHGs, nitrous oxide has the atmospheric lifetime of 120 year and global warming potential (GWP) is 310 kg equivalent CO_2 [3, 4]. And also it is expected as a dominant ozone depleting substance [5]. The feed water nitrogen amount will fix the N_2O emissions potential. Several possible treatments are available to produce N_2O, i.e. aerobic, anaerobic and combination of both. Number of research has been published on the nitrous oxide emissions from wastewater treatment plants. Still researchers are finding difficulties about the N_2O production mechanism in wastewater treatment plants. The main concerned gases in the wastewater treatment plants are CO_2, CH_4 and N_2O [6]. To evaluate the emissions, empirical static models and mechanistic models are available to determine the GHGs emissions to describe the performance of a wastewater treatment plants (WWTPs) [7]. N_2O emissions also has been considered in this model.

This paper considered N_2O production and operating conditions and also it deals with the carbon footprints of wastewater treatment plants. Direct GHG emissions are only considered in this work. The indirect emissions of GHGs like use of energy, biogas production, sludge disposal and the chemicals usage are the boundaries which are not considered in the evaluation of GHGs. N_2O production is evaluated in the denitrification process of WWTP, which is only covered in this study. The CO_2 and N_2O emissions play an important role in the WWTPs. So this paperwork will be focussing on the treatment of municipal sewage using SBR with respect to the carbon credit scheme.

A number of the research work has been done on the SBR for municipal wastewater treatment with respect to carbon credits. Some of the research works are discussed in Table 13.1. As per the literature, Chai et al. [6] concluded that SBR is a time-consumable process in wastewater treatment process which is similar to plug flow continuous reactors. But the results show the higher production of N_2O. In the carbon footprint evaluation, two types of assessments are available, i.e. high risk and low risk. This paper considered low risk in N_2O production. For low-risk evaluation, the average value of the data range is considered.

Neutral and negative in carbon footprint have been widely discussed and investigated. The energy contents in the wastewaters are in the form of COD and biosolids. The energy available in wastewater is less than the energy required. Neutrality possibility is a questionable one by the theoretical knowledge. A proper settling in the sedimentation tank with anaerobic digestion, addition of the carbon source, by reducing CO_2 emissions and energy generation from methane (by anaerobic) are the possible ways to get the neutral carbon footprint [7].

Table 13.1 Literature table

S. No.	List of papers	Parameters monitored (mg/L)						Removal efficiency (%) or (g/m^2 d)	
		pH	TSS	COD	BOD	Nitrogen	Phosphorus	CO_2	N_2O
1	Chai et al. [6]	7.4–7.9	2400	360–50	180–10	48–15	0.5	22–49	23–43
2	Trautmann et al. [9]	7.2–8.2	2500	920–270	210–80	60–35	NA	52	70
3	Linda et al. [10]	7.6	160–620	10 (COD/N)	0.16–0.48	75	NA	NA	NA
4	Jeon et al. [11]	6.8–7.5	3500	600	NA	40	15	NA	NA
5	Fang et al. [12]	7.8–8.5	401	300	NA	60	5.5	NA	NA
6	Pei et al. [13]	7.4	3000–3400	336	143	42	6.2	NA	NA
7	Marlies et al. [1]	6.5–6.8	NA	4.5 (COD/N)	NA	NA	NA	NA	58%
8	Gupta and Singh [14]	7–8	NA	4.2–5.8 (COD/N)	NA	23–35	NA	3027.84	298
9	Law et al. [15]	6.5	NA	>4 (COD/N)	NA	20–70	NA	NA	32–64%
10	Dong et al. [16]	7.6–8.2	1300	<400	<180	NA	NA	227.8	98%
11	Suresh Kumar and Sekaran [17]	NA	450	NA	NA	1–5%	NA	25–45%	NA
12	Mahvi [18]	8–12	6000	91%	98%	86%	82%	NA	NA
13	Hu [19]	7.1–8.3	550–110 (80%)	717–110 (84%)	193–5.1 (97%)	86.4–6.8 (92.2%)	27.2–12.6 (53.7%)	NA	NA
14	Wagner et al. [20]	8.2–8.7	630	87–92%	NA	69–88%	NA	NA	NA
15	Anupam et al. [21]	6.5–8.2	5500	1000–1100 (85–92%)	NA	40–90 (80%)	NA	NA	NA
16	Yu et al. [22]	6.7–7.4	2600	250–350	NA	37.6–52.7	5.4–7.6	NA	NA

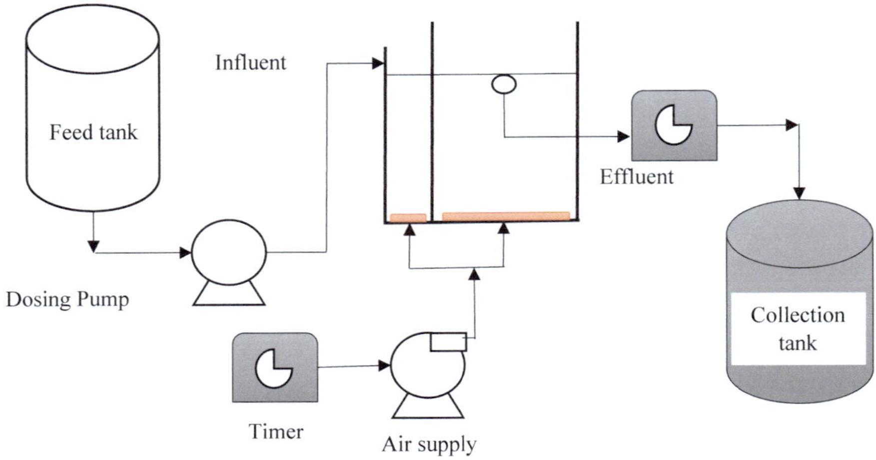

Fig. 13.1 Experimental set-up of sequencing batch reactor (pilot scale)

13.2 Materials and Methods

Analysis of various water characteristics for the collected sample from the SBR, test procedures for the analysis of sample and the estimation of greenhouse gases (GHGs) emissions were presented (Fig. 13.1).

13.3 Experimental Set-Up

A small-scale SBR reactor was designed to evaluate the nitrogen and phosphorus removal from the municipal wastewater treatment. The wastewater used as a feed that collected from the south-east zone of Puducherry. Reactor tank is equipped with aeration supply and suspended growth system. SBR tank is made up of plastic material which is having a dimension of 50 * 30 * 40 (length * breath * height) and a working volume of 60 L. Aeration supply was done by aquarium aeration pump. To feed the collected wastewater, peristaltic pump was used. PLC time controller was used to operate the pump and solenoid valve which are helpful in the proper operation and SBR processes like fill, react, settle and draw periods in the reactor. Samples were collected from the treated wastewater. Preliminary treatments are helpful to enhance the efficiency of the SBR. It includes screening and sedimentation. To achieve this effective removal, feed water is allowed for preliminary treatment such as to remove the grit, plastics, oil/grease, scum, etc. This is common for the most of the wastewater treatment plants. Sedimentation is used to settle the suspended solids. Preliminary treatments are healthy for pumps also. After the

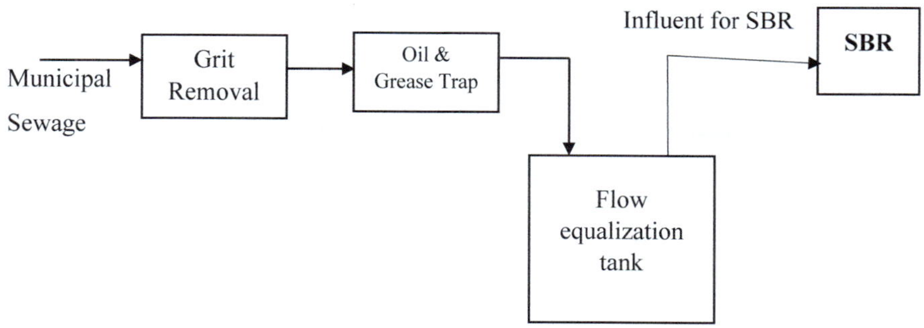

Fig. 13.2 Flow diagram of the overall process in the Municipal Sewage Treatment Plant

preliminary tests, the wastewater is allowed in flow equalization tank. For nitrification and denitrification, flow equalization is a mandate one. To maintain the food to microorganism ratio in wastewater treatment, flow equalization tank was used. Figure 13.2 shows the schematic representation of overall treatment process which includes preliminary treatment, sedimentation and flow equalization tank. After the flow equalization its all.

13.4 Sample Details

The samples were collected from influent and effluent of SBR in weekly four or three times simultaneously for a regular monitoring. During the running time of a reactor, grab samples are collected at a regular interval. After feeding the wastewater to the reactor, an initial sampling interval of 15 min and later longer interval of 30–45 min was followed. Especially in fill phase, a 7-min interval was maintained to find out the variations in the reactor. All the collected samples were analysed within 24 h from the sampling time for accuracy in results.

13.5 Analytical Methods for Water Characteristics

Collected samples were allowed for analysis with the standard methods of Central Public Health and Environmental Engineering Organization (CPHEEO) [8]. The parameters considered in this study are pH, electrical conductivity (EC), oxidation reduction potential (ORP), total dissolved solids (TDS), total suspended solids (TSS), volatile suspended solids (VSS), chemical oxygen demand (COD), biological oxygen demand ($BOD_{5,20\,°C}$) and total nitrogen (TN).

13.5.1 pH, EC and ORP

The characteristics of the sample collected from the SBR were analysed by various types of analytical methods. The pH, electrical conductivity (EC) and oxidation reduction potential (ORP) were analysed by using MP-6 Portable Meter. The specification of the metre was given in Table 13.2.

13.5.2 Total Solids (TS)

Total solids are the summation of dissolved and suspended solids. A known amount of sample is taken in a crucible and admitted in hot air oven for 103 ± 2 °C for a certain time. Initial crucible weight (W_1) and after evaporation of water with the remaining residues (W_2) are measured and calculated as per the following Eq. (3.1). The sum of increase in the weight of the dish and the weight of the dry dish is W_2 grams. Hence, the total solids are calculated as follows,

$$\text{Total Solids (mg/L)} = \left[(W_2 - W_1)/10^6\right]/\text{Volume of Sample} \qquad (3.1)$$

13.5.3 Total Dissolved Solids (TDS)

The procedure is same as that of total solids except that the sample is filtered and the residue along with W_1 is W_2. The weight loss of filtrate concentration is TDS. The filter used is glass microfiber filter (lum, Whatman GF/C). Equation (3.1) is used to estimate the concentration of total dissolved solids.

Table 13.2 Specifications of analytical metre

Company Name	Hach
pH (MP-6 Model)	0–14 pH
ORP (MP-6 Model)	(+ or −) 999 mV
Conductivity	0–999 μS/cm
	10–20 mS/cm
	in 5 auto ranges
TDS	0–999 PPM
	10–20 PPT
	in 5 auto ranges
Temperature	0–71 °C (32–100 F)

13.5.4 *Total Suspended Solids (TSS)*

The procedure is same as that of total dissolved solids except that the sample is filtered and the residue along with W_1 is W_2. The weight loss of residue concentration is TSS. The filter used is glass microfiber filter (lum, Whatman GF/C). Equation (3.1) is used to evaluate the concentration of TSS.

13.5.5 *Total Volatile Solids (TVS)*

The residue obtained from total solids is ignited at a temperature of 550 + 50 °C along with a dish in a muffle furnace and then weighed as W_3. Hence, the total volatile solids are calculated as follows,

$$\text{Total Volatile Solids (mg/L)} = \left[(W_2 - W_3)/10^6\right]/\text{Volume of Sample} \quad (3.2)$$

13.5.6 *Total Volatile Dissolved Solids (TVDS)*

After drying the filtrate in the oven for 1 h, placed in a muffle furnace at 550 °C for 20 min. The difference between the weights before and after the muffle furnace is the TVDS.

13.5.7 *Total Volatile Suspended Solids (TVSS)*

The residue remaining after drying at 103 °C is weighed and placed in a muffle furnace at 550 °C. The weight loss from ignition determines "volatile suspended solids". The difference between TVS and TVDS is TVSS.

13.5.8 *Chemical Oxygen Demand (COD)*

The measurement of all the organic compounds (biodegradable and non-biodegradable) in water is COD. The method is named as open tube reflux method. Mercury sulphate is initially added in the COD tubes to neutralize the effects of chlorides. Followed by pottasium di chromate and sulphuric acid were added in

Table 13.3 Sample volume range from Lovibond Oxidirect

Range BOD (mg/L)	Sample volume (mL)
0–40	428
0–80	360
0–200	244
0–400	157
0–800	94
0–2000	56
0–4000	21.7

COD tube. Silver sulphate is used as a catalyst. Ferrous ammonium sulphate was used as a titrant. COD was calculated by using the following Eq. (3.3):

$$\text{COD as mg/L} = (a - b) \times N \times 8000/\text{mL sample} \quad (3.3)$$

where

a mL FAS used for blank,
b mL FAS used for sample,
N normality of FAS.

13.5.9 Biochemical Oxygen Demand

BOD is the amount of oxygen needed to oxidize the organic matter that is present in the water. To measure BOD, Lovibond Oxidirect, Germany equipment was used. The samples were readjusted in pH for the range of 6.5 and 7.5. Lovibond BOD system works on the principle of pressure difference within a closed system which is called respirometric BOD. Initially, the measurement range was fixed and sample volume has taken according to Table 13.3 which is prescribed by Lovibond Oxidirect. Magnetic rod was inserted for a proper mixing. About 3–4 drops of KOH solution was placed in seal gasket. Bottles are screwed and placed inside the BOD incubator for 20 °C. SANCO BOD incubator was used to maintain the 20 °C. Readings were measured in the display monitor.

13.6 Estimation of CH_4 and CO_2 [2]

Aerobic processes mainly produce CO_2, whereas anaerobic process produces the combination of CH_4 and CO_2 [2]. The assumption was made that all organic carbon are removed from the wastewater. The estimation of CO_2 and CH_4 is calculated directly from the formula (3.4) and (3.5) [7]. The equations were adapted from “Greenhouse gas emissions estimation methodologies for Biogenic Emissions from selected categories” prepared by Research Triangle Institute (RTI).

Emission rate of CO_2 (mg of CO_2 per h) and CH_4 (mg of CH_4 per h) is as follows,

$$CO_2 = Q_{WW} \times Eff_{OD} \times CF_{CO_2} \times OD \times [(1 - MCF_{WW} * \ BG_{CH_4})(1 - \lambda)] \times 10^{-6} \quad (3.4)$$

$$CH_4 = Q_{WW} \times OD \times Eff_{OD} \times CF_{CH_4} \times \ [(MCF_{WW} \times BG_{CH_4})(1 - \lambda)] \times 10^{-6} \quad (3.5)$$

The biomass yield is the ratio of converted biomass to consumed biomass in grams. The biomass yield can be calculated by using Eq. (3.6).

$$\lambda = \frac{Q_S \times MLVSS_S \times CF_S}{Q_{WW} \times OD \times Eff_{OD} \times CF_C} \quad (3.6)$$

Now the estimation of CO_2 and CH_4 are possible by equation of (3.7) and (3.8).

$$CO_2\ (mg\ CO_2/h) = Q_s \times MLVSS \times CF_s \times (44/12) \times (1 - MCF_s \times BG_{CH_4}) \times 10^{-6} \quad (3.7)$$

$$CH_4\ (mg\ oCH_4/h) = Q_s \times MLVSS \times CF_s \times (16/12) \times (1 - MCF_s \times BG_{CH_4}) \times 10^{-6} \quad (3.8)$$

(Derived by Research Triangle Institute, NC, U.S. 2010).

where,

Q_{WW}	Influent flow rate (m^3 per h),
OD	Oxygen demand for the process which is explained as either BOD_5,
Eff_{OD}	Efficiency of oxygen demand removal,
CF_{CO_2}	CF for CO_2 production per unit of oxygen demand,
CF_{CH_4}	CF for CH_4 production per unit of oxygen demand,
MCF_{WW}	CF for methane in WWT unit (oxygen demand by anaerobic process),
BG_{CH_4}	Carbon fraction as methane for biogas generation (default is 0.65),
λ	Biomass yield,
Q_S	Stream flow rate of sludge (m^3 per h),
$MLVSS_S$	Mixed liquor volatile suspended solids (mg/L),
OD	Oxygen demand of influent wastewater to the biological treatment unit determined as either BOD5 or COD (mg/L = g/m^3),
Eff_{OD}	Effective oxygen demand removal,

Table 13.4 Correction factors values for Eqs. (3.4), (3.5) and (3.6)

Correction factor (CF)	Value	Unit
CO_2	44/32	CO_2/g oxygen demand
CH_4	16/32	CH_4/g oxygen demand
MLVSSs	0.53	g C/g oxygen demand
Carbon consumption	12/32	g C/g oxygen demand

CF_S CF for carbon (C) content (i.e. $MLVSS_S$),
CF_C CF for maximum carbon consumption (g C/g oxygen demand),
10^{-6} Units conversion factor (mg/g).

Methane correction factor (MCF) for the anaerobic sludge digestion is 0.8, and the biomass yield value λ can be obtained from wastewater treatment (Table 13.4).

13.7 Estimation of N_2O [2]

N_2O production in wastewater treatment plants depends upon the nitrogen present in the influent. To evaluate the N_2O production in wastewater treatment plants (N_2O_{WWTP} = mg N_2O/h), Eq. (3.9) was used. Nitrous oxide emission (EF_{N_2O}) factor unit is gram of nitrogen emitted as N_2O per gram of TKN in the feed.

$$N_2O_{WWTP} = Q_i \times TKN_i \times EF_{N_2O} \times (44\backslash 28) \times 10^{-6} \tag{3.9}$$

(Research Triangle Institute, NC, U.S. 2010).

where,

Q_i feed wastewater flow rate (m^3/h),
TKN_i Total Kjeldahl Nitrogen amount in the feed (mg/L).

13.8 Results and Discussion

The experimental results of both influent and effluent water samples that are collected from the sequencing batch reactor were presented and discussed in this paper. The presentations of results are explained in the following subdivisions to highlight the significance of salient details obtained from the experimental study. The changes in the water characteristics are critically assessed and presented the variations of pH, EC, TDS, TSS and COD, which were shown in Table 13.5.

Table 13.5 Analytical results of the sample collected: phase-I

S. No.	Date of sample collected	Sample	PH	EC	Temp	TDS	TSS	TVDS	TVSS	COD
		Units	–	(µS)	(°C)	(mg/L)	(mg/L)	(mg/L)	(mg/L)	(mg/L)
1	11-08-14	Inlet	8.03	1902	31.6	1130	120	410	85	350
		Outlet	8.53	1675	32	938	102	200	65	240
2	03-09-14	Inlet	7.45	1888	31.9	1033	97	440	70	320
		Outlet	7.25	1867	31.6	908	82	250	50	96
3	16-09-14	Inlet	7.37	1722	26.1	1134	246	230	140	300
		Outlet	6.94	1693	26.1	993	77	140	40	210
4	14-10-14	Inlet	7.36	1854	31.7	1710	252	280	110	392
		Outlet	7.4	1864	32.8	1650	210	160	130	130

13.9 Analytical Results for the Influent and Effluent of the SBR

The samples were collected from the influent and treated effluent of the sequencing batch reactor and analysed for various parameters. The samples were collected at regular intervals and detention time. The performance of SBR was analysed based on the detention time (Tables 13.6 and 13.7).

Table 13.6 Analytical results of the sample collected: phase-II

S. No.	Date	Sample collected	pH	EC	TDS	TSS	VSS	COD	BOD
			–	(µS)	(mg/L)	(mg/L)	(mg/L)	(mg/L)	(mg/L)
1	05-01-2015	Influent	6.36	1960	1370	90	460	320	–
		Effluent	6.38	1970	1360	120	130	130	–
2	16-02-2015	Influent	7.46	2168	990	70	540	300	180
		Effluent	7.79	2184	1990	140	160	110	85
3	24-02-2015	Influent	6.69	2124	2340	234	500	340	–
		Effluent	7.61	2190	2160	80	130	120	–
4	05-03-2015	Influent	7.43	1972	1355	120	80	90	180
		Effluent	6.74	1890	1296	258	170	300	96
5	09-03-2015	Influent	7.57	1786	1228	280	130	390	190
		Effluent	7.76	2054	1412	130	390	160	95
6	10-03-2015	Influent	7.08	1795	1235	252	150	360	180
		Effluent	7.36	2035	1405	110	160	110	85
7	11-03-2015	Influent	7.16	2082	1434	264	140	290	210
		Effluent	7.43	1972	1355	120	80	90	96
8	12-03-2015	Influent	6.74	1890	1296	258	170	300	190
		Effluent	7.54	2153	1482	100	80	80	95
9	13-03-2015	Influent	7.14	1968	1355	270	250	260	180
		Effluent	8.43	2160	1484	130	100	110	80

(continued)

Table 13.6 (continued)

S. No.	Date	Sample collected	pH	EC	TDS	TSS	VSS	COD	BOD
			–	(μS)	(mg/L)	(mg/L)	(mg/L)	(mg/L)	(mg/L)
10	16-03-2015	Influent	7.26	1794	1235	260	160	240	190
		Effluent	7.68	2060	1419	120	90	80	95
11	17-03-2015	Influent	7.36	1879	1288	260	180	290	180
		Effluent	7.78	2229	1534	110	440	90	96
12	18-03-2015	Influent	7.23	1864	1276	220	240	300	190
		Effluent	7.92	2310	1560	100	420	80	80
13	19-03-2015	Influent	7.38	1827	1260	270	260	260	190
		Effluent	7.42	2192	1555	130	450	110	95
14	20-03-2015	Influent	7.04	1666	1144	122	250	240	180
		Effluent	8.01	2314	1588	36	430	80	85
15	23-03-2015	Influent	7.30	1971	1344	126	270	240	210
		Effluent	7.51	2282	1568	32	420	80	96
16	24-03-2015	Influent	7.27	1651	1137	83	630	300	–
		Effluent	8.11	2163	1489	40	470	80	–
17	25-03-2015	Influent	7.15	1710	1175	195	290	260	180
		Effluent	7.97	2017	1388	52	280	110	80
18	26-03-2015	Influent	7.34	1638	1125	75	320	240	190
		Effluent	8.36	1975	1359	141	330	80	95
19	27-03-2015	Influent	7.12	1574	1080	30	310	290	–
		Effluent	8.12	2070	1420	150	330	90	–
20	30-03-2015	Influent	7.28	2148	1476	134	520	260	190
		Effluent	8.19	2119	1458	32	340	110	95
21	31-03-2015	Influent	7.12	1967	1354	126	310	300	180
		Effluent	8.44	2060	1418	48	340	120	85
22	01-04-2015	Influent	7.21	1927	1320	114	360	240	210
		Effluent	8.43	2208	1518	38	380	80	96

Table 13.7 Analytical results for the sludge present in the SBR tank

S. No.	Date	Sample collected	pH	EC	TDS	MLSS	SVI	TN
			–	(μS)	(mg/L)	(mg/L)	(mL/mg)	(mg/L)
1	24-03-2015	Reactor	7.89	2225	1531	2500	110	40
		Effluent	7.51	2282	1568	32		15
2	31-03-2015	Reactor	7.46	2156	1496	3200	75	38
		Effluent	8.44	2060	1418	48		12
3	08-04-2015	Reactor	7.56	2334	1610	3640	60	48
		Effluent	8.26	2196	1511	45		18
4	10-04-2015	Reactor	7.06	2352	1614	2430	85	45
		Effluent	7.94	1964	1349	52		15
5	15-04-2015	Reactor	7.34	2097	1442	4390	65	42
		Effluent	7.34	1768	1216	50		10

13.9.1 Removal Efficiency of SBR

Average TSS removal efficiency = 78%
Average COD removal efficiency = 65%
Average BOD removal efficiency = 52%
Average nitrogen removal efficiency = 67.8%.

The graphical representations give the variation of various parameters on the municipal sewage. The samples were collected from influent and effluent of the SBR tank and that were analysed using several analytical methods. Based on the analytical results, the graphs were plotted between the days (*X*-axis) and ranges (*Y*-axis) for the influent and effluent of the SBR tank. The removal efficiency of the SBR can be found using these graphical representations. For the total nitrogen analysis graph (Fig. 13.11), the samples were collected from the sludge present in the reactor tank and its effluent. Figures 13.3, 13.4, 13.5, 13.6, 13.7, 13.8, 13.9, 13.10, 13.11 and 13.12 are showed the effect of various parameters like pH, EC, TDS, TSS, TVS, COD, BOD, TN & removal efficiency respectively.

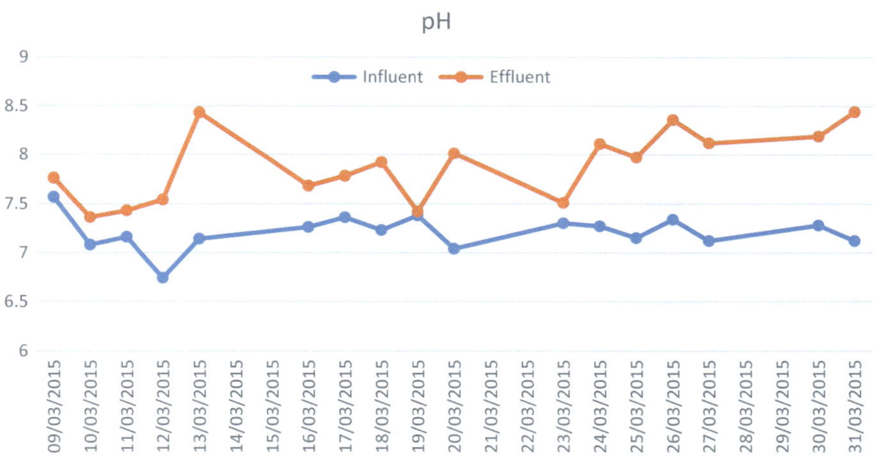

Fig. 13.3 Effect of pH values in both influent and effluent

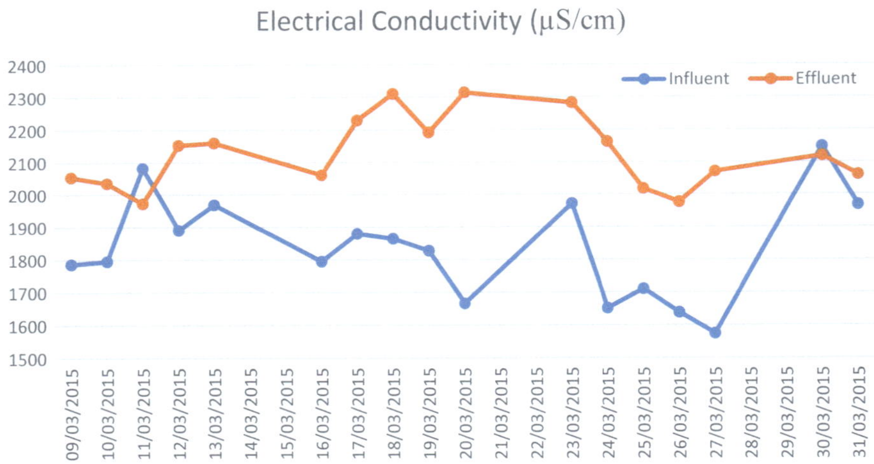

Fig. 13.4 Effect of electrical conductivity (EC) values in both influent and effluent

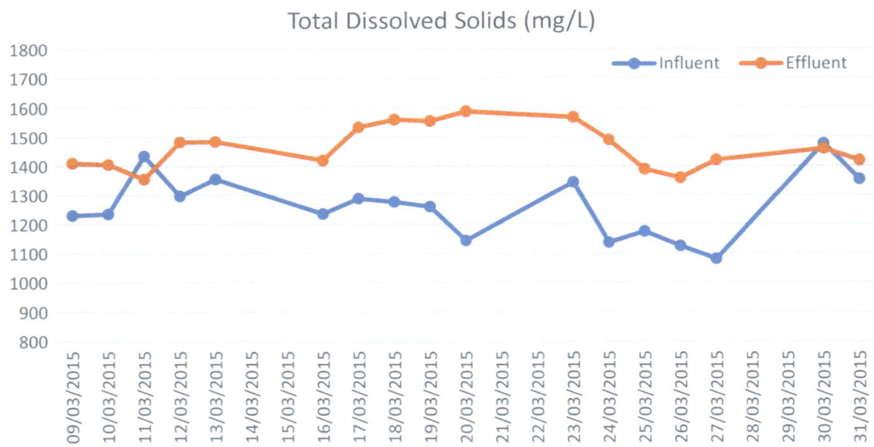

Fig. 13.5 Effect of total dissolved solids (TDS) values in both influent and effluent

13.10 Emission of CO_2 and N_2O

The emission of CO_2 was obtained in the wastewater treatments, mostly the CO_2 were raised their emission in the biological treatments of wastewater which is due to the biochemical reactions took place during the treatment of wastewater. The aeration which are given in the treatment systems are the main factors for the emission of CO_2 in the wastewater treatment systems. The CO_2 emission level is always

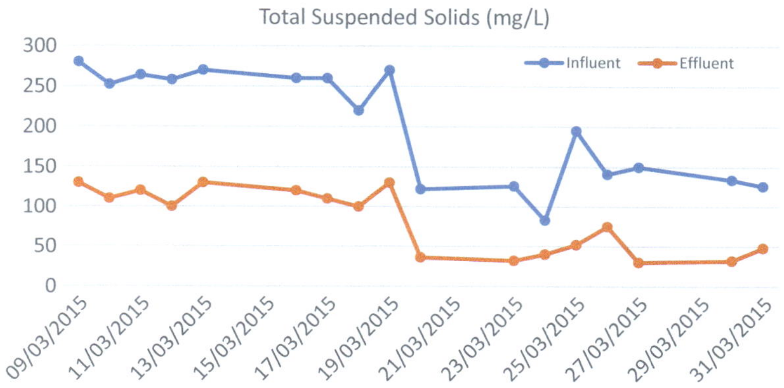

Fig. 13.6 Effect of total suspended solids (TSS) values in both influent and effluent

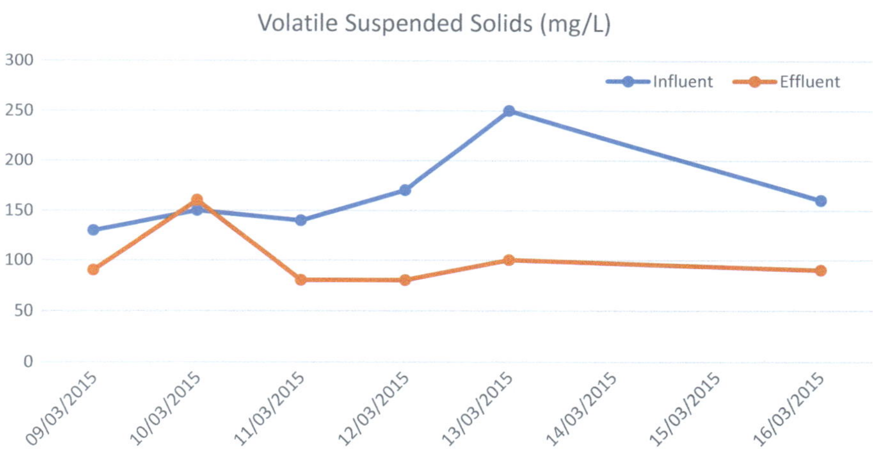

Fig. 13.7 Effect of volatile suspended solids (VSS) values in both influent and effluent

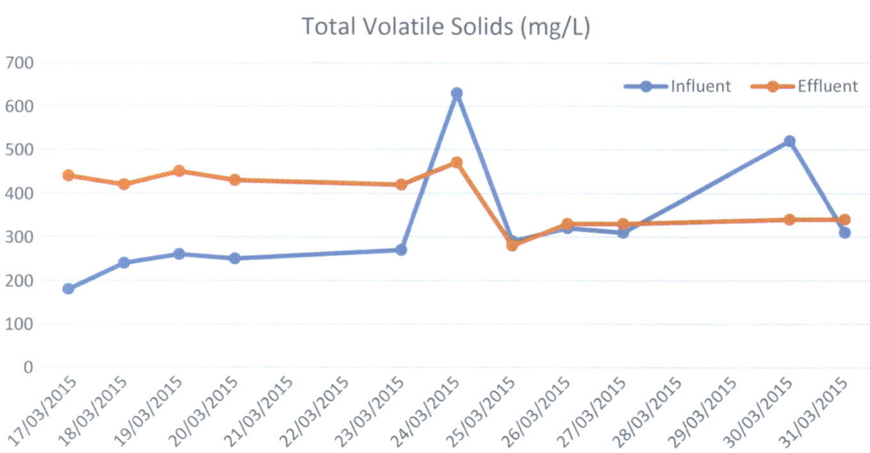

Fig. 13.8 Effect of total volatile solids (TVS) values in both influent and effluent

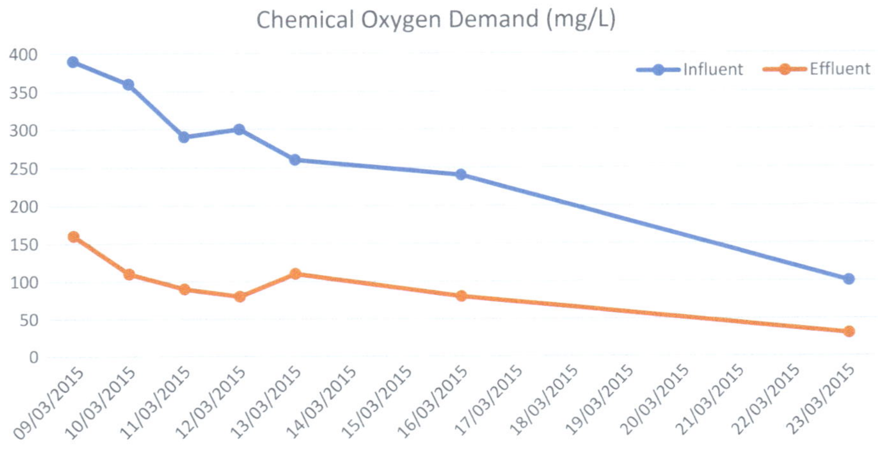

Fig. 13.9 Effect of COD values in both influent and effluent

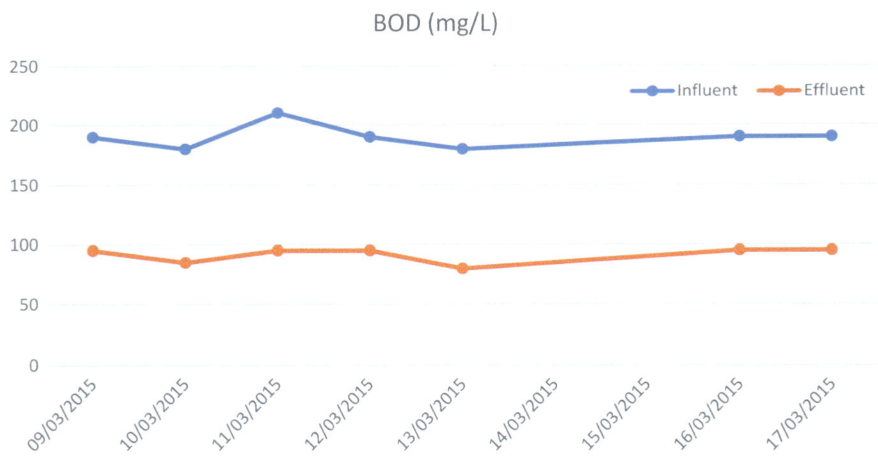

Fig. 13.10 Effect of BOD values in both influent and effluent

higher in all the treatment methods of wastewater when compared to other greenhouse gases. Here, the average CO_2 emission rate was determined as 0.01643 kg of CO_{2e} per year (Table 13.8). This determination was done by using the formula [7]. N_2O was formed during the aerated period and at the beginning of the non-aerated phase until the dissolved oxygen concentration dropped below 1.2 mg O_2/L. Then, N_2O concentration decreased until the aeration was restarted. These results indicate that N_2O was formed by AOB as a side product of ammonium oxidation and degraded biological (het. denitrification) during non-aerated periods (the decrease of N_2O is higher than theoretical emission rates) [9]. The nearly linear increase of the

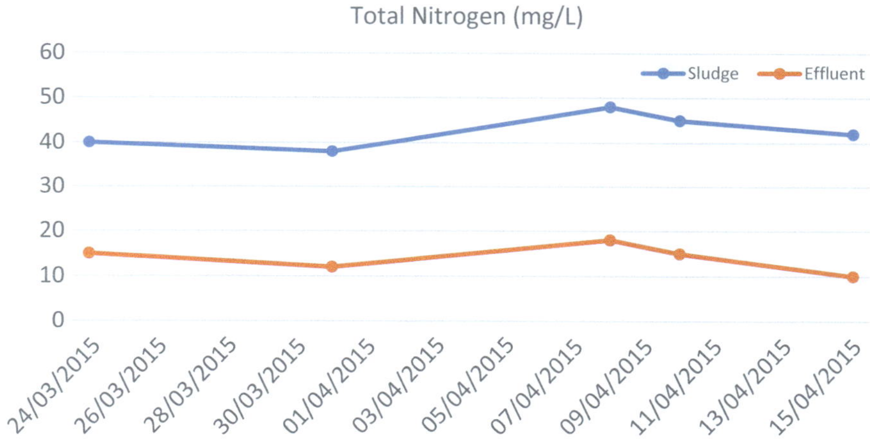

Fig. 13.11 Effect of total nitrogen (TN) values in both influent and effluent

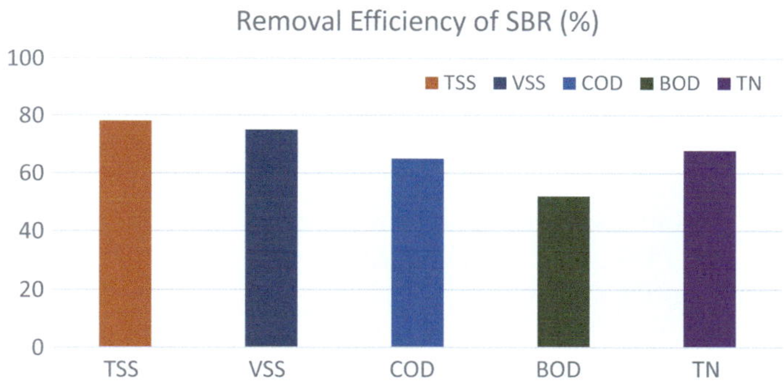

Fig. 13.12 Graphical representation on the removal efficiency of the SBR

N_2O concentration during the aerated phases despite intermittent aeration indicated that it is appropriate to assume a constant k_La-value in the emission calculations for the whole aerated phase. The average calculated N_2O emission rates were 0.005 kg N_2O–N/min (aerated) and 0.0003 kg N_2O–N/min (non-aerated). The resulting average N_2O emission factor is 0.74% N_2O–N/Nrem (0.52–1.5%), equalling 3.5 kg CO_2-eq/kg Nrem. Compared to values reported in the literature for nitrogen removal processes, this N_2O emission factor must be regarded rather low (e.g. 0.6–25.3%, Foley et al., 2010; 0.7–2.9%, Schneider et al., submitted) (Table 13.9).

In the treatment of municipal sewage using SBR, the nitrogen was removed in the reactor tank itself by the denitrification process obtained during the aeration period. Due to the nitrification and denitrification process, the N_2O is emitted on the treatment of wastewater. The N_2O is the most violent gas when compared to other greenhouse gases. Based on the total nitrogen (TN) values that are determined from the samples collected on the SBR tank and its effluent, the N_2O emission rate was

Table 13.8 Determined values of CO_2 emission rate in the SBR tank

S. No.	Date	MLVSS (mg/L)	SVI (mL/mg)	Emission rate of CO_2 (mg/day)
1	24-03-2015	2150	110	3.2931
2	31-03-2015	2950	75	4.523
3	08-04-2015	3400	60	5.2078
4	10-04-2015	2100	85	3.2165
5	15-04-2015	4100	65	6.2799

Table 13.9 Determined values of N_2O emission rate in the SBR tank

S. No.	Date	MLVSS (mg/L)	SVI (mL/mg)	Total nitrogen (TN) (mg/L)		Emission rate of N_2O (mg/day)
				Reactor tank	Effluent	
1	24-03-2015	2150	110	40	15	8.1005×10^{-4}
2	31-03-2015	2950	75	38	12	7.6953×10^{-4}
3	08-04-2015	3400	60	48	18	9.7205×10^{-4}
4	10-04-2015	2100	85	45	15	9.1130×10^{-4}
5	15-04-2015	4100	65	42	10	8.5053×10^{-4}

found using the formula which is mentioned above [7]. The determined N_2O emission rate is 3.1488×10^{-14} kg of N_2O per year. This is the average value of N_2O emission rate (Fig. 13.13).

In this study, the CO_2 footprint of a deammonification plant for part-stream treatment was calculated including the contribution of indirect (construction, energy) and direct emissions (N_2O). The results demonstrate that N_2O emissions make up the biggest share of total CO_2 emissions due to low specific reactor volume and energy demand for nitrogen removal and the high global warming potential of N_2O. Furthermore, N_2O emissions are the most sensitive factor within the CO_2 footprint requiring plant-specific measurements. The deammonification process for reject water treatment is an advantageous option to increase the nitrogen removal capacity of wastewater treatment plants compared to the effort (construction, energy and external carbon source addition) required to achieve combined treatment in the mainstream. In order to further reduce the CO_2-footprint of part-stream technologies, the development of innovative technologies for reduction of N_2O emissions is necessary. A possible approach is to combine the prevention of N_2O formation with minimization of gas transfer and N_2O degradation [9] that is currently investigated.

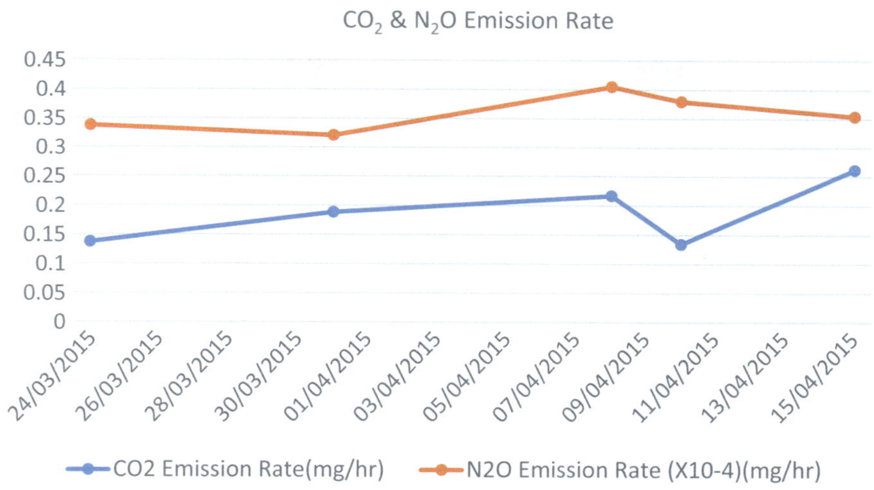

Fig. 13.13 Effect on the emission rate of CO_2 and N_2O from the SBR

13.11 Neutral Carbon Footprint Measure

Design and operating condition of SBR will be helpful in the neutral carbon footprint. Design includes long solid retention time, a large volume of reactor and high MLR rate. Chai et al. [6], mentioned appropriate process is anammox, i.e. it leads to the less energy consumption.

13.12 Limitations

This experimental study gives the concern in neutral carbon footprint and evaluation of GHG emissions with respect to carbon credits. Still, it has limitations and needs further research to enhance the mentioned objective. The pilot-scale SBR capacity is less (0.06 m^3) and the real influent is 2.578 m^3/day. The scale variation may collapse the accuracy results. Operating studies also need to be considered with the incorporation of neutral carbon footprint.

13.13 Conclusion

From the results and discussion, it is clear that the performance of SBR technology for the treatment of municipal WWT was better than any other treatment methods. And it is very effective in the reduction of greenhouse gases (GHGs) such as N_2O in the sewage treatment system. SBR has the pros of less volume consumption of tank; solid retention time can be controllable; ease of sampling, etc. Design and process conditions are helpful to reduce the carbon footprints. The evaluated carbon footprint for this SBR is 0.01643 kg CO_{2e} per year. From SBR, the main GHG emissions are CO_2 (from aerobic treatment), N_2O (direct emissions) and electricity use (indirect emissions). In the current study, N_2O emissions need to be taken care. For example, incomplete denitrification may cause the high N_2O generation. Operating conditions and designing of SBR need to give more concentration for the better results, and it concluded that it needs more detailed research and onsite measurement also needed to be included in the future research.

References

1. Kampschreur MJ, Temmink H, Kleerebezem R, Jetten MS, van Loosdrecht MC (2009) Nitrous oxide emissions during wastewater treatment. Water Res 43(17):4093–4103
2. Peavy HS, Rowe DR, Tchbanoglous G (1985) Environmental engineering. McGraw-Hill International Editions. Civil Engineering Series, pp 207–322
3. Intergovernmental Panel on Climate Change. Internet site: http://www.ipcc.ch. Accessed on 10 Dec 2016
4. Greenhouse Gas and Global Warming Potential Values. Excerpt from the inventory of US Greenhouse Emissions and Sinks: 1999–2000 (2002). Internet site: www.epa.gov/globalwarming/publications/emissions. Accessed on 11 Dec 2016
5. Wang Y, Geng J, Guo G, Wang C, Liu S (2011) N_2O production in anaerobic/anoxic denitrifying phosphorous removal process: the effects of carbon sources shock. Chem Eng J 172(2–3):999–1007
6. Chai C, Zhang D, Yu Y, Feng Y, Wong MS (2015) Carbon footprint analyses of mainstream wastewater treatment technologies under different sludge treatment scenarios in China. Water 7:918–938
7. Research Triangle Institute (RTI) International (2010) Greenhouse gas emissions estimation methodologies for biogenic emissions from selected source categories: solid waste disposal wastewater treatment ethanol fermentation, pp 1–43
8. Ministry of Urban Development, Govt of India. Central Public Health & Environmental Engineering Oraganization. Guide manual: Water and Wastewater Analysis. Website: http://cpheeo.nic.in/. Accessed on 12 Jan 2015
9. Trautmann N, Schneider Y, Vogel B, Beier M, Rosenwinkel KH (2013) Determination of the carbon footprint for nitrogen removal from part-streams including indirect and direct emissions—a case study. Institute for Sanitary Engineering and Waste Management, Leibniz Universität Hannover, Welfengarten 1, 30167 Hannover, Germany
10. Figueroa LA, Miller J, Dawson HE (1997) Biodegradation of two polyethoxylated non-ionic surfactants in sequencing batch reactors. Water Environ Res 69(7):1282–1289
11. Ok JC, Lee DS, Lee MW, Park JM (2001) Enhanced biological phosphorus removal in an anaerobic-aerobic sequencing batch reactor: effect of pH. Water Environ Res 73(3):301–306
12. Fang HHP, Member, ASCE, Liu Y (2000) Intracellular polymers in aerobic sludge of sequencing batch reactors. J Environ Eng 126(8):732–738
13. Pei H, Hu W, Pan J, Zhang J (2008) Simultaneous nitrogen and phosphorus removal using a double sludge switching sequencing batch reactor. J Environ Eng 134(11):923–927
14. Diksha G, Singh SK (2012) Greenhouse gas emissions from wastewater treatment plants: a case study of Noida. J Water Sustain 2(2):131–139

15. Fudala-Ksiazek S, Luczkiewicz A, Fitobor K, Olanczuk-Neyman K (2014) Nitrogen removal via the nitrite pathway during wastewater co-treatment with ammonia-rich landfill leachates in a sequencing batch reactor. Environ Sci Pollut Res 21:7307–7318
16. Tai PD, Li PJ, Sun TH, He YW, Zhou QX, Gong ZQ, Motoyuki M (2002) Greenhouse gas emissions from a constructed wetland for municipal sewage treatment. J Environ Sci 14 (1):27–33
17. Suresh Kumar D, Sekaran V (2013) Clean development mechanism in sewage treatment plants. IOSR J Environ Sci, Toxicol Food Technol (IOSR-JESTFT) 3(2):36–44
18. Mahvi AH (2008) Sequencing batch reactor: a promising technology in wastewater treatment. Iran J Environ Health Sci Eng 5(2):79–90
19. Hu JY, Ong SL, Ng WJ, Liu W (2005) Use of a sequencing batch reactor for nitrogen and phosphorus removal from municipal wastewater. J Environ Eng 131(5):734–744
20. Wagner J, da Costa RHR (2013) Aerobic granulation in a sequencing batch reactor using real domestic wastewater. J Environ Eng 139(11):1391–1396
21. Anupam D, Mukherjee S, Datta S (2006) Sequencing batch reactor (SBR) treatment for simultaneous organic carbon and nitrogen removal—a laboratory study. J Environ Sci Eng 478(43):169–174
22. Yu RF, Liaw SL, Cheng WY, Chang CN (2000) Performance enhancement of SBR applying Real-time control. J Environ Eng 126(10):943–948

Part VI
Integrated System

Chapter 14
Convective Heat Transfer of Metal Oxide-Based Nanofluids in a Shell and Tube Heat Exchanger

Nishant Kumar and Shriram S. Sonawane

Abstract Nanofluid is a solid–liquid mixture in which metallic or nonmetallic nanoparticles are suspended in the base fluid. The convective heat transfer performance for CuO- and TiO_2-based nanofluids was measured flowing in the tube side in shell and tube heat exchanger. The effect of CuO and TiO_2 nanoparticles on the overall heat transfer coefficient of base fluid like distilled water was studied. Nanofluids showed an enhancement in the overall heat transfer coefficient. The investigation of thermal conductivity and heat transfer coefficient enhancement was analyzed with different concentration of nanoparticles, base fluids, sonication time, and temperature of fluids. The nanoparticles concentration was 0.01–0.06 vol% used in base fluids. The heat transfer performance was studied for different Peclet number and temperature of nanofluids. An increment in the heat transfer performance was found for the nanofluids, by increasing the concentration of nanoparticles, flow rate and temperature of nanofluid.

Keywords CuO/TiO_2 nanofluids · Convective heat transfer · Heat exchanger Nusselt number · Reynolds number · Peclet number · Volume fraction

Nomenclature

$\emptyset$	Volume fraction
δ_{f}	Density of fluid
δ_{p}	Density of particle
δ_{nf}	Density of nanofluid
C_{Pp}	Specific heat of particle
C_{Pf}	Specific heat of fluid
C_{Pnf}	Specific heat of nanofluid
q	Convective heat transfer
A	Area

N. Kumar · S.S. Sonawane (✉)
Department of Chemical Engineering, Visvesvaraya National Institute of Technology, Nagpur 440010, Maharashtra, India
e-mail: shriramsonawane@gmail.com

S. Kumar et al. (eds.), *Conference Proceedings of the Second International Conference on Recent Advances in Bioenergy Research*, Springer Proceedings in Energy, https://doi.org/10.1007/978-981-10-6107-3_14

Nu Nusselt number (Dimensionless)
Re Reynolds number (Dimensionless)
Pr Prandtl number (Dimensionless)
K Thermal conductivity

14.1 Introduction

The industrial processes involve the transfer of heat energy. The consumption of energy becomes an important phenomenon in different industrial process because of the limitation of fossil fuel. The heat exchanger is the very important instrument related to transfer of heat and energy. Various efforts have been taken to enhance heat transfer, reduce the time of the heat transfer, and increase the efficiency of fluids. A number of experiments have been done on the heat transfer performance for their various applications to heat transfer enhancement [1].

The fluids which have poor heat transfer characteristics act as a barrier for different types of heat exchanger. The thermal conductivity of solid particles is higher than that of common base fluids that is the reason to disperse solid particles in the fluids to better thermal conductivity and heat transfer rate. The difficulty of using bigger size particles such as micro-sized particles in the fluid medium are the poor suspensions and clogging inside the flow channels.

In modern technology, nanometer-sized particles are used, which have better thermal properties than micro-sized particles. It provides opportunities to process and produce materials with average crystallite sizes below 100 nm [2]. The fluids which contain the nanometer-sized particles are nanofluids. Nanometer-sized particles give better suspension in base fluids, and it also nullifies the effect of clogging [3]. Nanoparticles suspension in fluids can alter the heat transfer characteristics of base fluids. Nanoparticles have more surface area than the conventional particles, and it significantly improves the heat transfer characteristics and also increases the stability of the suspension [4]. Keblinski et al. [5] studied and discuss the properties of nanofluids and its future challenges. There are many techniques to increase the heat transfer performance such as: by enhancing thermal conductivity of the fluid, changing the flow geometry or boundary conditions [6]. Commonly used heat transfer fluids are water, ethylene glycol, or paraffin, which has low thermal conductivities than solid particles.

Several existing published articles focused on enhancing thermal conductivity and heat transfer. The experimental results showed the enhancement of thermal conductivity and heat transfer by the addition of nanoparticles. Arani et al. [7] conducted the convective heat transfer study of TiO_2/water nanofluid in fully developed flow with different nanoparticles and volumetric concentration (0.01–0.02%) were analyzed. It was found that the Nusselt number of nanofluids was higher than the base fluids. Heris et al. [8] analyzed the heat transfer characteristics of Al_2O_3-based nanofluid in a circular tube with constant wall temperature. The authors found that the heat transfer coefficient increases as the concentration of nanoparticles increases in the base fluid. Kim et al. [9] studied the effect of the heat transfer coefficients of thermal conductivity. It was found that the thermal conductivity has an important role in the improvement of the convective heat transfer coefficients. Sharma et al. [10] studied the convective heat transfer of Al_2O_3/water nanofluid with constant heat flux and found that the heat transfer coefficient increases when increasing Reynolds number. Rohit et al. [11, 12] analyzed the thermal conductivity and overall heat transfer coefficient of Al_2O_3-based nanofluids. It was found that 3 vol% nanofluids showed most favorable performance with overall heat transfer coefficient 16% more than water.

In this paper, the convective heat transfer coefficient of nanofluid in a shell and tube heat exchanger under steady state and turbulent flow conditions has been studied experimentally. The effects of Peclet number and particle volume concentration on heat transfer are analyzed. Furthermore, the experimental results are compared with the proposed empirical model, and validation of the method is investigated.

14.2 Materials and Methods

14.2.1 Nanofluid Preparation

The nanofluid preparation was done by two-step method; initially, nanoparticles were prepared separately and then mixed it in the base fluid with the help of ultra sonication. The two nanoparticles (CuO and TiO_2) and distilled water were used as a base fluid for the preparation of nanofluids. Nanoparticles concentration was varied from 0.01 to 0.06% volume concentration with 26 and 9 nm average diameters of CuO and TiO_2 nanoparticles, respectively. The equivalent weight of nanoparticles was measured and added to the base fluid [13]. The suspension was then sonicated for a different sonication time starting from 40 to 90 min in a sonicator. Physical properties of both the nanoparticles are shown in Table 14.1.

14.3 Experimental Setup

CuO- and TiO_2-based nanofluids were used to flow through the tube side of shell and tube heat exchanger. The heat transfer performance was measured for both the nanofluids at different flow rate and different concentration of nanoparticles. Hot

Table 14.1 Physical properties of the nanoparticles

Nanoparticles	Mean diameter (nm)	Density (kg/m^3)	Specific heat (kJ/kg K)	Thermal conductivity (W/m K)
CuO	26	6310	540	18
TiO_2	9	3900	710	8.4

Table 14.2 specification of heat exchanger

Length of tubes	0.6 m
Number of tubes	8
Diameter of tubes (Outer)	0.0137 m
Diameter of tubes (Inner)	0.0107

water flows through the shell side, and the nanofluids flow through the tube. Peristaltic pumps are there controlling the flow rate of fluids. Temperature sensors are provided for temperature measurement of cold fluid and hot fluid. The specification of the heat exchanger is shown in Table 14.2.

14.4 Data Processing

Nusselt number and Prandtl number of nanofluids with various particle volume concentrations and flow rate were calculated to analyze the enhancement in convective heat transfer.

The heat transfer rate of the nanofluid is

$$q = mC_{\mathrm{eff}}(T_{\mathrm{out}} - T_{\mathrm{in}}) \tag{14.1}$$

where q is the rate of heat governed in the hot water, m is the mass flow rate of nanofluids, and T_{out} and T_{in} are the outlet and inlet temperatures of the nanofluid. C_{eff} is the effective specific heat of the nanofluid which can be calculated from Xuan and Roetzel relation [3] (Table 14.3).

$$C_{\mathrm{Pnf}} = (1 - \emptyset)\delta_{\mathrm{f}} C_{\mathrm{Pf}} + \emptyset(\delta_{\mathrm{p}} C_{\mathrm{Pp}}) \tag{14.2}$$

The density of the nanofluids is calculated by

$$\delta_{\mathrm{nf}} = (1 - \emptyset)\delta_{\mathrm{f}} + \delta_{\mathrm{p}}\emptyset \tag{14.3}$$

The overall heat transfer coefficient, U_0, is calculated by

$$U_0 = \frac{q}{A_0}\mathrm{LMTD} \tag{14.4}$$

Table 14.3 Experimental conditions

Parameter	Hot fluid	Cold fluids
Concentration of nanoparticles	0	0.01–0.06
Temperature	50–80 (°C)	30 (°C)
Flow rate	1 LPM	0.5–3 LPM
Reynolds number	1000–3000	1000–5000

14.5 Results and Discussions

The system was first tested with the distilled water for the accuracy of the measurements before performing the convective heat transfer experiments of nanofluids. Figures 14.1 and 14.2 represent the value of the overall heat transfer coefficient versus the Peclet number at a different flow rate of nanofluids and different concentration of nanoparticles. The results clearly show the increment in overall heat transfer coefficient as the concentration of nanoparticles and the flow rate of nanofluids increase. As the constant Peclet number, the overall heat transfer coefficient for both nanofluids increased as the concentration of nanoparticles increases in the base fluid. The maximum enhancement found in CuO/water nanofluids at 0.06% volume concentration and the Peclet number of about 46,000 are approximately 19% shown in Fig. 14.1. And the enhancement at the same Peclet number (46,000) for all three different volume concentrations of nanoparticles (0.02, 0.04, and 0.06%) is about 10, 14, and 19%, respectively. In the case of TiO_2/water nanofluid, the maximum enhancement found at 0.06% volume concentration was 12% for Peclet number 45,000 shown in Fig. 14.2. At a certain Peclet number (45,000), the enhancement of the overall heat transfer coefficient was found 6, 9, and 12% for 0.02, 0.04, and 0.06 vol%, respectively.

Figures 14.3 and 14.4 illustrate the enhancement of convective heat transfer coefficients versus Peclet number of different concentrations of CuO and TiO_2 nanoparticles in the nanofluids, respectively. As it is clearly shown in Figs., the addition of nanoparticles in the fluid increases the convective heat transfer coefficient of both nanofluids. The maximum enhancement of the convective heat transfer coefficients was found at maximum concentration (0.6 vol%) of CuO/water and TiO_2/water nanofluids. This enhancement for 0.2, 0.4, and 0.6 vol% of

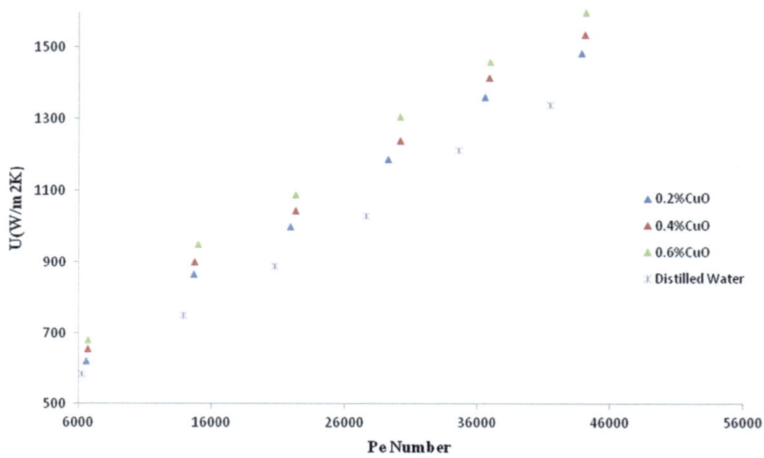

Fig. 14.1 Overall heat transfer coefficient of CuO/water nanofluid versus Peclet number for various volume concentrations

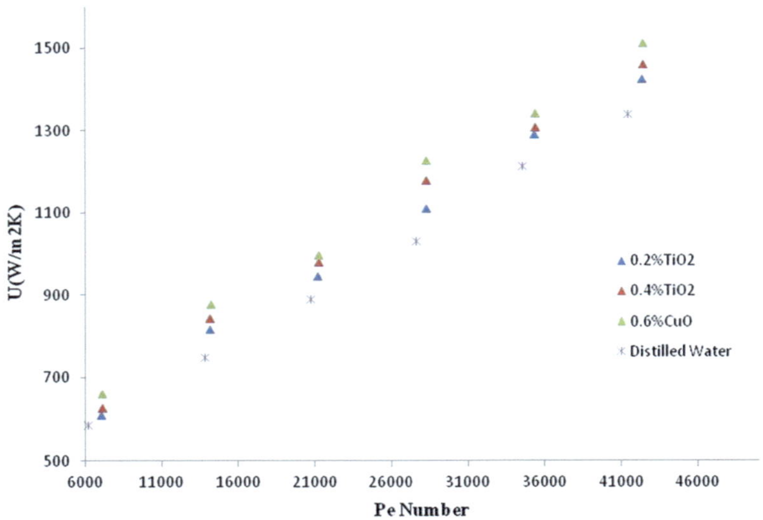

Fig. 14.2 Overall heat transfer coefficient of TiO_2/water nanofluid versus Peclet number for various volume concentrations

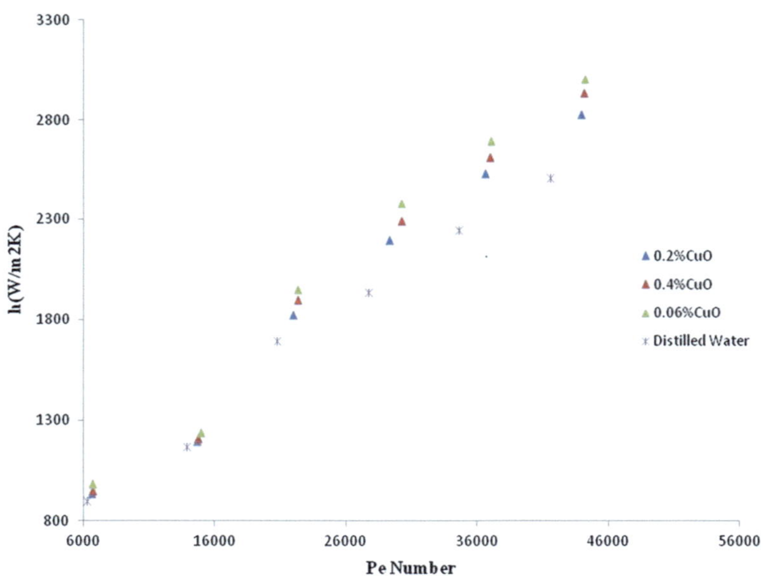

Fig. 14.3 Convective heat transfer coefficient of CuO/water nanofluid versus Peclet number for different volume concentrations

CuO/water at maximum flow rate of nanofluids are about 12, 17, and 27%, respectively. In the case of TiO_2/water nanofluid, the enhancement occurred 10, 14, and 18% in 0.2, 0.4, and 0.6 Vol% of TiO_2 nanoparticles in base-fluid, respectively.

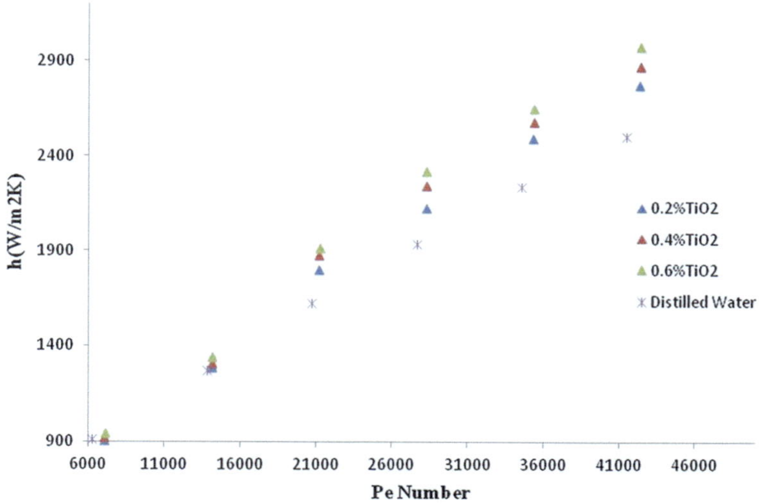

Fig. 14.4 Convective heat transfer coefficient of TiO_2/water nanofluid versus Peclet number for different volume concentrations

The thermal conductivity of fluid and thermal boundary-layer thickness is the reason for the enhancement of convective heat transfer coefficient. The boundary-layer thickness decreases when the thermal conductivity of the fluid increases because of the mobility of nanoparticles near the wall, it reduces the viscosity at the wall region, and thermal boundary-layer thickness decreases. As it is clearly shown that the enhancement in the convective heat transfer coefficient of CuO/water is more than that of TiO_2/water nanofluid. The difference may be because of the thermal conductivity of CuO nanoparticles is more than that of TiO_2 nanoparticles.

14.6 Comparison Between Experimental Results and Available Correlations

The validation of experimental results for both nanofluids is done by comparison with the correlations of Dittus-Boelter correlation [14]. Xuan and Li correlation [14] and Gnielinski correlation [1], given in Eqs. 14.5, 14.6, and 14.7.

$$Nu = 0.023 Re^{0.8} Pr^{0.3} \quad \text{Dittus} - \text{Boelter} \tag{14.5}$$

$$Nu = 00059\left(1 + 7.6286\emptyset^{0.6886} Pr^{0.0001}\right) Re^{0.9238} Pr^{0.4} \quad \text{Xuan and Li} \tag{14.6}$$

$$Nu = 0.012(Re^{0.87} - 280)Pr^{0.3} \quad \text{Gnielinski correlation} \tag{14.7}$$

Peclet number, Reynolds number and the Nusselt number for nanofluid are defined respectively as:

$$Re = \frac{\rho v d}{\mu} \tag{14.8}$$

$$Nu = \frac{hd}{K} \tag{14.9}$$

$$Pr = \frac{C_p \mu}{K} \tag{14.10}$$

$$Pe = Re * Pr \tag{14.11}$$

The comparison between the experimental and theoretical results for both the nanofluids is shown in Figs. 14.5 and 14.6, respectively.

Results indicate that a good agreement exists between the experimental results, and the theoretical values by all the three correlations were provided for laminar to turbulent flow reason. All three correlations are given higher Nusselt numbers in comparison with the experimental Nusselt number at the maximum flow rate (3 LPM). The Nusselt number values for the Gnielinski correlation indicate the most suitable model in the turbulent in comparison with other two correlations. At the concentration of 0.06 vol% of CuO nanoparticles at Reynold's number 7793, the deviation is 8% for the Gnielinski correlation and for the TiO_2 nanoparticles at 0.06 vol%, and Reynolds number 6988 shows 34% deviation with the experimental result.

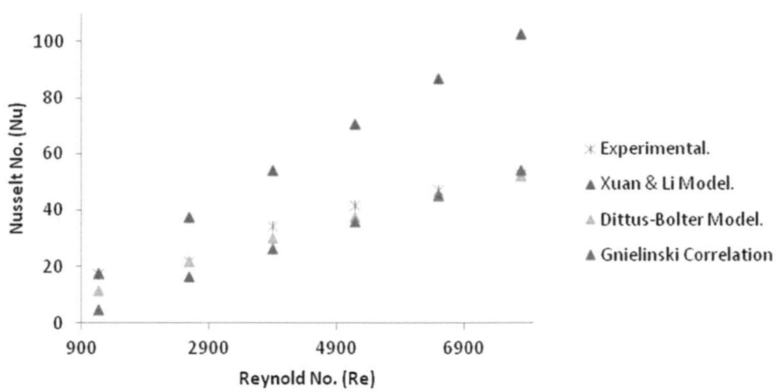

Fig. 14.5 Experimental Nusselt number of CuO/water nanofluid versus Reynolds number at 50 °C hot fluid 0.06 vol% concentration

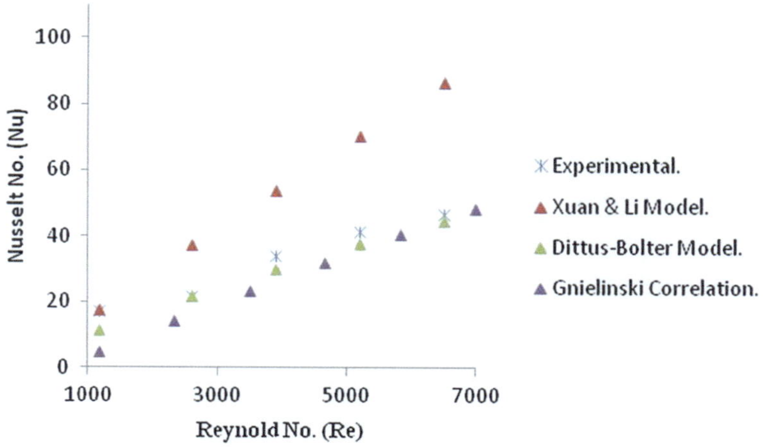

Fig. 14.6 Experimental Nusselt number of TiO_2/water nanofluid versus Reynolds number at 50 °C hot fluid 0.06 vol% concentration

14.7 Conclusion

- Convective heat transfer for both the nanolfuids was studied for a wide range of Reynolds number (from laminar to turbulent flow region).
- An enhancement in the overall heat transfer coefficient has been studied for both the nanofluids for different concentration, flow rate, and temperature of hot fluids.
- The experimental study showed that the heat transfer performance of both the nanofluids enhanced with Reynolds number.
- The convective heat transfer increases as the addition of nanoparticles increases in the base fluid, it is because the thermal conductivity of nanoparticles is more than the fluid. So, it creates a thermal bridge between the nanoparticles and fluid. Hence, more heat transfer takes place for the nanofluid.

Acknowledgements The authors are thankful to the Department of Science and Technology (DST) for their funding to this research project (No ETA/318/2012).

References

1. Farajollahi B, Etemad SG, Hojjat M (2010) Heat transfer of nanofluids in a shell and tube heat exchanger. Int J Heat Mass Transf 53:12–17
2. Dilek EF (2009) Preparation of nano-fluid and determination of thermal conductivity. Master's Thesis, Atatürk University

3. Xuan Y, Roetzel W (2000) Conceptions for heat transfer correlation of nanofluids. Int J Heat Mass Transf 43:3701–3707
4. Sahin B, Gultekin GG, Manay E (2013) Experimental investigation of heat transfer and pressure drop characteristics of Al_2O_3-water nanofluid. Exp Therm Fluid Sci 50:21–28
5. Keblinski P, Eastman JA, Cahill DG (2005) Nanofluids for thermal transport. Mater Today 8 (6):36–44
6. Wang XQ, Mujumdar AS (2007) Heat transfer characteristics of nanofluids: a review. Int J Therm Sci 46:1–19
7. Arani AA, Amani J (2013) Experimental investigation of diameter effect on heat transfer performance and drop of TiO_2–water nanofluid. Exp Therm Fluid Sci 44:520–533
8. Heris SZ, Esfahany MN, Etemad SG (2007) Experimental investigation of convective heat transfer of Al_2O_3/water nanofluid in circular tube. Int J Heat Fluid Flow 28(2):203–210
9. Kim D, Kwon Y, Cho Y, Li C, Cheong S, Hwang Y, Lee J, Hong D, Moon S (2009) Convective heat transfer characteristics of nanofluids under laminar and turbulent flow conditions. Curr Appl Phys 9(2, Supplement 1) e119–e123
10. Sharma KV, Syam Sundar L, Sarma PK (2009) Estimation of heat transfer coefficient and friction factor in the transition flow with low volume concentration of Al_2O_3 nanofluid flowing in a circular tube and with twisted tape insert. Int Commun Heat Mass Transf 36:503–507
11. Khedkar RS, Sonawane SS, Wasewar KL (2012) Influence of CuO nanoparticles in enhancing the thermal conductivity of water and monoethylene glycol based nanofluids. Int Commun Heat Mass Transf 39(13):1306–1334
12. Khedkar RS, Sonawane SS, Wasewar KL (2013) Water to nanofluids heat transfer in concentric tube heat exchanger: experimental study (NUiCONE 2012). Procedia Eng 51:318–323
13. Das SK, Putra N, Thiesen P, Roetzel W (2003) Temperature dependence of thermal conductivity enhancement for nanofluids, transactions of ASME. J Heat Transf 125:567–574
14. Xuan Y, Li Q (2003) Investigation convective heat transfer and flow features of nanofluids. J Heat Transf 125(1):151–155

MIX
Papier aus verantwortungsvollen Quellen
Paper from responsible sources
FSC® C105338
FSC
www.fsc.org

If you have any concerns about our products,
you can contact us on
ProductSafety@springernature.com

In case Publisher is established outside the EU,
the EU authorized representative is:
Springer Nature Customer Service Center GmbH
Europaplatz 3, 69115 Heidelberg, Germany

Printed by Libri Plureos GmbH
in Hamburg, Germany